T0235041

Damage
Mechanics

MECHANICAL ENGINEERING
A Series of Textbooks and Reference Books

Founding Editor

L. L. Faulkner

*Columbus Division, Battelle Memorial Institute
and Department of Mechanical Engineering
The Ohio State University
Columbus, Ohio*

Damage Mechanics

George Z. Voyiadjis
Peter I. Kattan

CRC Press
Taylor & Francis Group
Boca Raton London New York

CRC Press is an imprint of the
Taylor & Francis Group, an **informa** business

CRC Press
Taylor & Francis Group
6000 Broken Sound Parkway NW, Suite 300
Boca Raton, FL 33487-2742

First issued in paperback 2019

ISBN-13: 978-0-8247-2757-4 (hbk)
ISBN-13: 978-0-367-39257-4 (pbk)

Library of Congress Cataloging-in-Publication Data

Voyiadjis, G. Z.
    Damage mechanics / George Z. Voyiadjis, Peter I. Kattan.
        p. cm. -- (Mechanical engineering)
    Includes bibliographical references and index.
    ISBN 0-8247-2757-6
    1. Continuum damage mechanics. I. Kattan, Peter Issa, 1961- II. Title. III. Mechanical engineering (Taylor & Francis)

TA409.V69 2005
620.1'123--dc22   2005041839

Visit the Taylor & Francis Web site at
http://www.taylorandfrancis.com

and the CRC Press Web site at
http://www.crcpress.com

Preface

The book presents the principles of continuum damage mechanics including the latest research by the authors. The presentation is theoretical in nature emphasizing the detailed derivations of the various models and formulations. The work of various active researchers in this area is also presented. Both isotropic and anisotropic damage mechanics theories are discussed. Also, both elastic and elasto-plastic damage analyses are presented. The presentation used in this book is consistent and systematic. Many examples illustrating the theory are presented especially in the early chapters of the book. In addition, a large number of problems appear at the ends of sections and chapters for students to be used as practice. The book can be used as a graduate textbook for students in the areas of civil engineering, mechanical engineering, aerospace engineering, materials science, and engineering mechanics. The book can also be used as a reference for active researchers in this field as well as for practicing engineers.

Chapter 1 provides the mathematical basis needed to understand the other chapters. This consists of the basics of tensor algebra and analysis. This chapter includes a large number of problems at the end of each section in order to emphasize the importance of this basic chapter for the book. In addition, the authors emphasize the use of the computer algebra system MAPLE in this chapter to solve problems. For this purpose, a short MAPLE tutorial is included and the various sections in the chapter include the needed MAPLE commands. In Chapter 2, a review is made of the basics of the theory of elasticity within the framework of continuum mechanics. Deformation, strains, and stresses are defined and analyzed in this chapter. The principles of continuum damage mechanics are introduced in Chapter 3 as the simple case of isotropic damage is discussed. In particular, the concept of effective stress is presented for the first time in this chapter. In addition, a section on damage evolution appears in this chapter. The kinematic description of damage is described in detail in Chapter 4. Chapter 5 continues with the principles of damage mechanics for the general case of anisotropic damage. The general definition of the effective stress is introduced here as well as a representation of the damage effect tensor. In chapter 6, a review of the theory of plasticity is presented emphasizing kinematic hardening. A constitutive model for damage plasticity is presented in Chapter 7. This involves formulating a coupled elasto-plastic damage theory that was developed by the authors. Finally, the kinematics of damage for finite-strain elasto-plastic solids is presented in Chapter 8. The book concludes with a comprehensive reference list.

The text increases gradually in difficulty from the basics in the first few chapters to the advanced mathematical and mechanical models in the later chapters. It is written in such a manner that the reader can progress from beginning to end in a very smooth way. Typically, students of damage mechanics are required to have finished the required prerequisites for this topic before reading this book. These prerequisites include tensor algebra, elasticity and plasticity. For those students with the necessary prerequisites, chapters 1, 2, and 6 can be skipped. However, if the students did not finish the necessary prerequisites, then they must read these three chapters (1, 2, and 6) before embarking on a study of damage mechanics. Therefore, this book is comprehensive in the sense that all the prerequisites needed for damage mechanics are reviewed in the book.

The presentation is limited to damage mechanics of metals and homogeneous materials. In case the student wishes to apply damage mechanics to metal matrix composites, then he is advised to read a previous book by the authors, entitled "*Advances in Damage Mechanics: Metals and Metal Matrix Composites*" by Voyiadjis and Kattan. The presentation in the book is also limited to theoretical aspects of damage mechanics. If the student wishes to study the numerical implementation of damage mechanics using finite elements, then he is referred to a previous book by the authors, entitled "*Damage Mechanics with Finite Elements: Practical Applications with Computer Tools*" by Kattan and Voyiadjis.

Finally, the authors would like to express their thanks and appreciation to the editors at Taylor and Francis (CRC Press) for providing the opportunity to publish this book in this form. In addition, the authors wish to thank their family members without whose help and support this book would not have appeared. The second author would like to acknowledge the financial support provided by the Center for Computation and Technology at Louisiana State University, headed by Dr. Edward Seidel.

George Z. Voyiadjis
voyiadjis@eng.lsu.edu
Peter I. Kattan
pkattan@lsu.edu

Authors

Dr. George Z. Voyiadjis, Ph.D.

George Z. Voyiadjis is the Boyd Professor at Louisiana State University in the Department of Civil and Environmental Engineering. This is the highest professorial rank awarded by the Louisiana State University System. He joined the faculty of Louisiana State University in 1980. Voyiadjis' primary research interest is in damage mechanics of metals, metal matrix composites, and ceramics with emphasis on the theoretical modeling, numerical simulation of material behavior, and experimental correlation. Research activities of particular interest encompass macromechanical/micromechanical constitutive modeling, experimental procedure for quantification of crack densities, inelastic behavior, thermal effects, interfaces, damage, failure, fracture, and numerical modeling. His experience also includes work on modeling of cyclic plasticity for metals. He has over 160 referred journal articles and 13 books (9 as editor) to his credit. Over 41 graduate students (21 Ph.D.) completed their degrees under his direction. He has also supervised 11 postdoctoral associates. Voyiadjis has been funded as a principal investigator from the National Science Foundation, the Department of Defense, the Air Force Office of Scientific Research, the Department of Transportation, and major companies such as IBM, and Martin Marietta. He has also been invited to give theme presentations and lectures in many countries around the world and to serve as guest editor in numerous volumes of the *Journal of Computer Methods in Applied Mechanics and Engineering, International Journal of Plasticity, Journal of Engineering Mechanics of the ASCE*, and *Jounal of Mechanics of Materials*. These special issues focus on the areas of damage mechanics, structures, fracture mechanics, localization, and bridging of length scales. Dr. Voyiadjis also had a two year stint in industry as a senior engineer with Nuclear Power services , Inc. and Ebasco Services Inc. During that period he was engaged in the research and development of stress analysis of nuclear power plants and was also involved in the development of finite element computer codes in conjunction with the piping analysis of power plants. He is currently a Fellow in the American Society of Civil Engineers, the American Society of Mechanical Engineers, and the American Academy of Mechanics.

Dr. Peter I. Kattan, Ph.D.

Peter I. Kattan has a Ph.D. in Civil Engineering from Louisiana State University. He has written three books on damage mechanics, one book on finite elements, and one book on composite materials. His research work is currently focused on damage mechanics with fabric tensors and the physical characterization of micro-crack distribution and their evolution. He has published extensively on theory of plates and shells, constitutive modeling of inelastic materials and damage mechanics. He is currently a Visiting Professor at Louisiana State University in Baton Rouge, Louisiana.

Contents

1

Mathematical Preliminaries

The mathematical background needed to study damage mechanics is introduced in this chapter. In particular, vectors and tensors are introduced using the computer algebra system Maple. A short Maple tutorial is given first followed by vector and tensor analysis. It should be noted that the coverage of vectors and tensors is not comprehensive – only the needed material is shown. For more details, the reader is referred to the references by Chung (1996), Eringen (1980), Fung (1965), Fung (1994), Green and Zerna (1968), Hjelmstad (1997), Lai et al. (1984), Lubliner (1990), Maugin (1992), Marsden and Hughes (1983), Lemaitre (1996), McDonald (1996), Mroz (1973), Segel (1987), and Werde (1972).

1.1 Maple Tutorial

In this section a short tutorial on using the computer algebra system Maple is given. For more details the reader is referred to the Maple books by Corless (1995), Heck (1996), Nicolaides and Walkington (1996), Kofler (1997), Schwartz (1999), and Cornil and Testud (2001). In addition, a search for Maple on the internet will reveal several Maple tutorials which can be freely downloaded.

Once you start Maple on your computer system, you can start using it immediately as a calculator as follows:

```
>   3+1/2;
```

$$\frac{7}{2}$$

Note that each Maple command is terminated by a semicolon. Note also that the result is given in exact form (in the form of a fraction in the example above). In case a numerical result is needed, you can use the Maple function *evalf* as follows:

```
> evalf(3+1/2);
```
$$3.500000000$$

Maple can be used easily to perform all sorts of algebraic manipulations like simplifying, expanding, and factoring. The following examples illustrate this:

```
> expand((x+2)^3);
```
$$x^3 + 6\,x^2 + 12\,x + 8$$

```
> simplify(sin(x)^2 + cos(x)^2);
```
$$1$$

Maple can also be used to solve algebraic equations as follows:

```
> solve(2*x^3 +5*x^2 -x -4,x);
```
$$-1, \; -\frac{3}{4} + \frac{1}{4}\sqrt{41}, \; -\frac{3}{4} - \frac{1}{4}\sqrt{41}$$

Systems of simultaneous algebraic equations can be solved numerically using the Maple command *fsolve* as follows:

```
> fsolve({x^2 -3*y^2 +6 = 0, sin(x)*sqrt(y) = 2},{x,y});
```
$$\{x = 7.497717372, \; y = 4.553964053\}$$

Differential and integral calculus can be performed using the Maple commands *diff* and *int* as follows:

```
> diff(x^5 - 2*x^3 + x^2 + 10, x);
```
$$5\,x^4 - 6\,x^2 + 2\,x$$

```
> int(2*cos(5*x)^3,x=a..b);
```
$$\frac{2}{15}\cos(5\,b)^2\sin(5\,b) + \frac{4}{15}\sin(5\,b) - \frac{2}{15}\cos(5\,a)^2\sin(5\,a) - \frac{4}{15}\sin(5\,a)$$

Simple ordinary differential equations can also be solved using the Maple command *dsolve* as follows:

```
> deq:=diff(y(x),x)*y(x)*(1-x^2) = 5*x - 1;
```

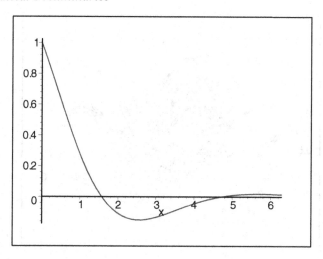

FIGURE 1.1
Two-Dimensional Plot Using Maple

$$deq := (\tfrac{\partial}{\partial x}\, y(x))\, y(x)\, (1 - x^2) = 5\,x - 1$$

```
> dsolve({deq,y(0)=0},y(x));
```

$$y(x) = \sqrt{-2\,\mathrm{arctanh}(x) + 5\,I\,\pi - 5\ln(-1 + x^2)},$$
$$y(x) = -\sqrt{-2\,\mathrm{arctanh}(x) + 5\,I\,\pi - 5\ln(-1 + x^2)}$$

Finally, two-dimensional and three-dimensional graphs can be plotted using the Maple commands *plot* and *plot3d* as follows:

```
> plot(cos(x)*exp(1)^(-2*x/3),x=0..2*Pi);
```

```
> plot3d(sin(x)*cos(y),x=0..2*Pi,y=0..2*Pi,axes=boxed);
```

See Figures 1.1 and 1.2 for the resulting plots of the above two commands, respectively.

Problems

1.1 Perform the following calculations using Maple. Obtain an exact result first followed by a numerical answer.

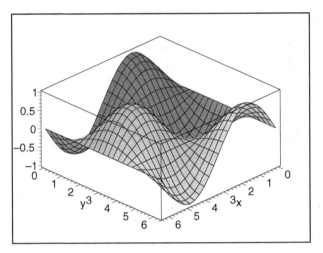

FIGURE 1.2
Three-Dimensional Plot Using Maple

    (a) $\sqrt{788}$

    (b) $\cos(\frac{\Pi}{2})$

    (c) $\frac{1}{\frac{1}{2}+\frac{1}{3}}$

    (d) $5^7 + e^3$

1.2 Perform the following algebraic manipulations using Maple.

    (a) Expand the expression $(2x - 1)^5$

    (b) Factor the expression $\cos^4 x - \sin^4 x$

    (c) Simplify the expression $1 - \sin^2 x - \cos^2 x$

1.3 Solve each one of the following equations using Maple.

    (a) $x^2 - 2x + 1 = 0$

    (b) $x^4 + x^3 - 2x^2 - x + 1 = 0$

    (c) $y^2 e^y = 5$

    (d) $\sin(2x)\cos(2x) = 1$

1.4 Obtain a numerical solution to each of the following sets of simultaneous algebraic equations using Maple.

    (a) $x + 2y = 5$; $3x - 5y = 4$

    (b) $x^2 + y^2 = 20$; $y * \sin(x) = 2$

1.5 Determine the specified derivative for each given expression using Maple.

(a) $\frac{dy}{dx}$ for $y = 3x^5 - 2x^2 + x + 1$

(b) $\frac{dy^2}{dx^2}$ for $y = e^x \sin(2x)$

(c) $\frac{dy}{dx}$ for $x + 3\sqrt{y} = 15$

1.6 Determine $\int f(x)dx$ for each $f(x)$ given below using Maple.

(a) $f(x) = x^2 + 2x + 1$

(b) $f(x) = \cos^2(4x) + \sin(2x)$

(c) $f(x) = \frac{x}{x^2+1}$

1.7 Solve each of the following ordinary differential equations using Maple.

(a) $2\frac{dy}{dx} + y + 5 = 0,$ $y(0) = 0$

(b) $x^3 y \frac{dy}{dx} = 2x^2 + x + 1,$ $y(1) = 1$

1.8 Plot a graph for each of the following functions using Maple.

(a) $\sin(\frac{x}{2}) + 2\cos(\frac{x}{3}), 0 \le x \le 6\Pi$

(b) $\sin(\frac{x}{2}) + 2\cos(\frac{y}{3}), 0 \le x \le 4\Pi; 0 \le y \le 6\Pi$

1.2   Vectors

In this section we explore vectors and vector operations using Maple. In order to use vectors in Maple, we need first to invoke the linear algebra package using the following command:

```
>  with(linalg);
```

Warning, the protected names norm and trace have been redefined and unprotected

[*BlockDiagonal*, *GramSchmidt*, *JordanBlock*, *LUdecomp*, *QRdecomp*, *Wronskian*, a*
addrow, *adj*, *adjoint*, *angle*, *augment*, *backsub*, *band*, *basis*, *bezout*, *blockmatrix*,
charmat, *charpoly*, *cholesky*, *col*, *coldim*, *colspace*, *colspan*, *companion*, *concat*,
cond, *copyinto*, *crossprod*, *curl*, *definite*, *delcols*, *delrows*, *det*, *diag*, *diverge*,
dotprod, *eigenvals*, *eigenvalues*, *eigenvectors*, *eigenvects*, *entermatrix*, *equal*,
exponential, *extend*, *ffgausselim*, *fibonacci*, *forwardsub*, *frobenius*, *gausselim*,
gaussjord, *geneqns*, *genmatrix*, *grad*, *hadamard*, *hermite*, *hessian*, *hilbert*,
htranspose, *ihermite*, *indexfunc*, *innerprod*, *intbasis*, *inverse*, *ismith*, *issimilar*,
iszero, *jacobian*, *jordan*, *kernel*, *laplacian*, *leastsqrs*, *linsolve*, *matadd*, *matrix*,
minor, *minpoly*, *mulcol*, *mulrow*, *multiply*, *norm*, *normalize*, *nullspace*, *orthog*,
permanent, *pivot*, *potential*, *randmatrix*, *randvector*, *rank*, *ratform*, *row*, *rowdim*,
rowspace, *rowspan*, *rref*, *scalarmul*, *singularvals*, *smith*, *stackmatrix*, *submatrix*,
subvector, *sumbasis*, *swapcol*, *swaprow*, *sylvester*, *toeplitz*, *trace*, *transpose*,
vandermonde, *vecpotent*, *vectdim*, *vector*, *wronskian*]

Figure 1.3 shows a two-dimensional vector **A** where **A**=(2,3). This vector
can be defined in Maple as follows:

```
>  A:=vector([2,3]);
```
$$A := [2, 3]$$

A three-dimensional vector **B** where **B**=(1,3,-4) is shown in Figure 1.4.
This vector can be defined in Maple as follows:

```
>  B:=vector([1,3,-4]);
```
$$B := [1, 3, -4]$$

Vectors can also be defined in Maple using variables or symbols as follows:

```
>  v1:=vector([x,y,z]);
```
$$v1 := [x, y, z]$$
```
>  v2:=vector([2*a,0,b-1]);
```
$$v2 := [2\,a, 0, b - 1]$$

The Maple command *evalm* can be used to perform some simple vector
operations as follows:

```
>  v:=vector([1,0,2]);
```

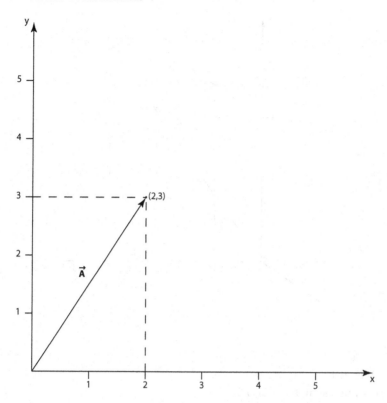

FIGURE 1.3
A two-dimensional vector $\overrightarrow{A}$

$$v := [1,\, 0,\, 2]$$

```
> w:=evalm(v+4);
```

$$w := [5,\, 4,\, 6]$$

The above command will add 4 to each element of the vector **v**. Scalar multiplication can also be performed as follows:

```
> y:=evalm(3*v);
```

$$y := [3,\, 0,\, 6]$$

Alternatively, the special Maple command *scalarmul* can be used to perform scalar multiplication as follows:

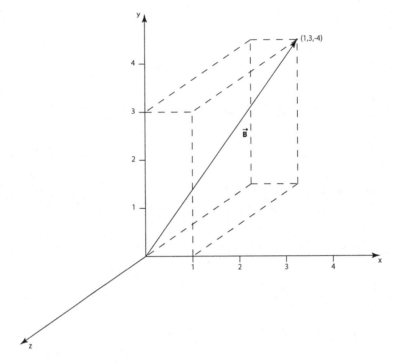

FIGURE 1.4

A three-dimensional vector $\vec{B}$

```
>  y:=scalarmul(v,3);
```
$$y := [3,\, 0,\, 6]$$

The command *evalm* can be used to perform various operations on vectors like vector addition and subtraction as follows:

```
>  a:=vector([x,0,2]);
```
$$a := [x,\, 0,\, 2]$$
```
>  b:=vector([1,1,y]);
```
$$b := [1,\, 1,\, y]$$
```
>  c:=evalm(a+b);
```
$$c := [x+1,\, 1,\, 2+y]$$
```
>  d:=evalm(a-b);
```
$$d := [x-1,\, -1,\, 2-y]$$
```
>  e:=evalm(2*a+3*b-c);
```

$$e := [x + 2,\ 2,\ 2 + 2\,y]$$

Vector products like the dot product and cross product can be performed on two vectors using the Maple commands *dotprod* and *crossprod* as follows:

```
> v:=vector([x,x^2,x^3]);
```
$$v := [x,\ x^2,\ x^3]$$
```
> w:=vector([1,2,3]);
```
$$w := [1,\ 2,\ 3]$$
```
> u:=dotprod(v,w);
```
$$u := x + 2\,x^2 + 3\,x^3$$
```
> y:=crossprod(v,w);
```
$$y := [3\,x^2 - 2\,x^3,\ x^3 - 3\,x,\ 2\,x - x^2]$$

The norm of a vector can be calculated using the Maple command *norm.* Several norms can be calculated according to the following formulae for a three-dimensional vector $\mathbf{v} = (v_1, v_2, v_3)$

$$\|\mathbf{v}\|_\infty = max(|v_1|, |v_2|, |v_3|) \tag{1.1a}$$

$$\|\mathbf{v}\|_1 = |v_1| + |v_2| + |v_3| \tag{1.1b}$$

$$\|\mathbf{v}\|_2 = \sqrt{(v_1)^2 + (v_2)^2 + (v_3)^2} \tag{1.1c}$$

$$\|\mathbf{v}\|_3 = \sqrt[3]{(v_1)^3 + (v_2)^3 + (v_3)^3} \tag{1.1d}$$

In general, the n-norm $\|\mathbf{v}\|_n$ is defined by the formula:

$$\|\mathbf{v}\|_n = \sqrt[n]{(v_1)^n + (v_2)^n + (v_3)^n} \tag{1.2}$$

The *norm* command has two arguments, *norm(*$\mathbf{v}$*,n)* where $\mathbf{v}$ is the vector and n=1,2,3,... is the order of the norm. The infinity-norm $\|\mathbf{v}\|_\infty$ is obtained by using the *norm* command with one argument only, *norm(*$\mathbf{v}$*)*. The following examples illustrate the use of the *norm* command:

```
> v:=vector([1,2,3]);
```
$$v := [1,\ 2,\ 3]$$
```
> norm(v);
```
$$3$$
```
> norm(v,1);
```
$$6$$

```
>   norm(v,2);
```
$$\sqrt{14}$$

```
>   norm(v,3);
```
$$36^{(1/3)}$$

```
>   evalf(norm(v,3));
```
$$3.301927249$$

The command *normalize* is used to normalize a vector to a unit vector of length 1 using the following formula for a three-dimensional vector $\mathbf{v} = (v_1, v_2, v_3)$:

$$\mathbf{v}_{\text{normalized}} = (\frac{v_1}{\|v\|_2}, \frac{v_2}{\|v\|_2}, \frac{v_3}{\|v\|_2}) \tag{1.3}$$

The *normalize* command is illustrated by the following example:

```
>   v:=vector([2,3,7]);
```
$$v := [2, 3, 7]$$

```
>   normalize(v);
```
$$\left[\frac{1}{31} \sqrt{62}, \frac{3}{62} \sqrt{62}, \frac{7}{62} \sqrt{62} \right]$$

It should be noted that the length of a vector is equal to its 2-norm as follows:

$$|\mathbf{v}| = \|\mathbf{v}\|_2 \tag{1.4}$$

The angle θ between two vectors $\mathbf{u}$ and $\mathbf{v}$ is calculated using the formula:

$$\cos\theta = \frac{\mathbf{u} \cdot \mathbf{v}}{\|\mathbf{u}\|_2 \|\mathbf{v}\|_2} \tag{1.5}$$

In Maple the command *angle* can be used to calculate the angle between two vectors. The resulting angle is given in radians - multiply it by $\frac{180}{\pi}$ to convert it into degrees as in the following example:

```
>   u:=vector([1,1,1]);
```
$$u := [1, 1, 1]$$

```
>   v:=vector([1,0,2]);
```
$$v := [1, 0, 2]$$

```
>   theta:=angle(u,v);
```
$$\theta := \arccos(\frac{1}{5} \sqrt{3} \sqrt{5})$$

```
>  theta:=evalf(theta);
```
$$\theta := .6847192024$$
```
>  theta:=evalf(theta*180/Pi);
```
$$\theta := 39.23152043$$

Maple can deal with general vectors in terms of their components even if these components are not specified. We can define two vectors **u** and **v** each with three unspecified components as follows:

```
>  u:=vector(3);
```
$$u := \text{array}(1..3, [])$$
```
>  v:=vector(3);
```
$$v := \text{array}(1..3, [])$$

The above commands define general three-dimensional vectors **u** and **v** with their components (u_1, u_2, u_3) and (v_1, v_2, v_3) unspecified. All the previous operations in this section can be applied to these vectors as shown in the following examples:

```
>  evalm(u+v);
```
$$[u_1 + v_1, \, u_2 + v_2, \, u_3 + v_3]$$
```
>  evalm(2*u-3*v);
```
$$[2\,u_1 - 3\,v_1, \, 2\,u_2 - 3\,v_2, \, 2\,u_3 - 3\,v_3]$$
```
>  scalarmul(v,3);
```
$$[3\,v_1, \, 3\,v_2, \, 3\,v_3]$$
```
>  dotprod(u,v);
```
$$u_1\,\overline{(v_1)} + u_2\,\overline{(v_2)} + u_3\,\overline{(v_3)}$$
```
>  crossprod(u,v);
```
$$[u_2\,v_3 - u_3\,v_2, \, u_3\,v_1 - u_1\,v_3, \, u_1\,v_2 - u_2\,v_1]$$
```
>  norm(v,2);
```
$$\sqrt{|v_1|^2 + |v_2|^2 + |v_3|^2}$$
```
>  normalize(u);
```
$$\left[\frac{u_1}{\sqrt{\%1}}, \, \frac{u_2}{\sqrt{\%1}}, \, \frac{u_3}{\sqrt{\%1}}\right]$$
$$\%1 := |u_1|^2 + |u_2|^2 + |u_3|^2$$
```
>  angle(u,v);
```

$$\arccos\left(\frac{u_1\,v_1 + u_2\,v_2 + u_3\,v_3}{\sqrt{u_1{}^2 + u_2{}^2 + u_3{}^2}\,\sqrt{v_1{}^2 + v_2{}^2 + v_3{}^2}}\right)$$

Defining general vectors in Maple is very helpful especially in deriving complicated equations involving vectors.

Example 1.1

Consider two three-dimensional vectors **u** and **v** . Use Maple to show that

$$|\,\mathbf{u} - \mathbf{v}\,|^2 = |\,\mathbf{u}\,|^2 + |\,\mathbf{v}\,|^2 - 2\,|\,\mathbf{u}\,||\,\mathbf{v}\,|\cos\theta \tag{1.6}$$

where θ is the angle between the two vectors.

Solution

Use general vectors to define **u** and **v** as three-dimensional vectors in Maple as follows:

```
>   u:=vector(3);
```
$$u := \operatorname{array}(1..3,\ [])$$
```
>   v:=vector(3);
```
$$v := \operatorname{array}(1..3,\ [])$$

Next, evaluate the left-hand side using Maple as follows:

```
>   w:=evalm(u-v);
```
$$w := [u_1 - v_1,\ u_2 - v_2,\ u_3 - v_3]$$
```
>   x:=norm(w,2);
```
$$x := \sqrt{|-u_1 + v_1|^2 + |-u_2 + v_2|^2 + |-u_3 + v_3|^2}$$
```
>   left_side:=x^2;
```
$$left\_side := |-u_1 + v_1|^2 + |-u_2 + v_2|^2 + |-u_3 + v_3|^2$$
```
>   left_side:=simplify(left_side, symbolic);
```
$$left\_side := u_1{}^2 + u_2{}^2 + u_3{}^2 + v_1{}^2 + v_2{}^2 + v_3{}^2 - 2\,u_1\,v_1 - 2\,u_2\,v_2 - 2\,u_3\,v_3$$

Finally, evaluate the right-hand side using Maple as follows:

```
>   y:=norm(u,2);
```
$$y := \sqrt{|u_1|^2 + |u_2|^2 + |u_3|^2}$$

```
>   z:=norm(v,2);
```

$$z := \sqrt{|v_1|^2 + |v_2|^2 + |v_3|^2}$$

```
>   theta:=angle(u,v);
```

$$\theta := \arccos\left(\frac{u_1 v_1 + u_2 v_2 + u_3 v_3}{\sqrt{u_1{}^2 + u_2{}^2 + u_3{}^2}\,\sqrt{v_1{}^2 + v_2{}^2 + v_3{}^2}}\right)$$

```
>   right_side:=y^2 + z^2 -2*y*z*cos(theta);
```

$$right\_side := |u_1|^2 + |u_2|^2 + |u_3|^2 + |v_1|^2 + |v_2|^2 + |v_3|^2$$
$$- \frac{2\sqrt{|u_1|^2 + |u_2|^2 + |u_3|^2}\,\sqrt{|v_1|^2 + |v_2|^2 + |v_3|^2}\,(u_1 v_1 + u_2 v_2 + u_3 v_3)}{\sqrt{u_1{}^2 + u_2{}^2 + u_3{}^2}\,\sqrt{v_1{}^2 + v_2{}^2 + v_3{}^2}}$$

```
>   right_side:=simplify(right_side, symbolic);
```

$$right\_side := u_1{}^2 + u_2{}^2 + u_3{}^2 + v_1{}^2 + v_2{}^2 + v_3{}^2 - 2\,u_1 v_1 - 2\,u_2 v_2 - 2\,u_3 v_3$$

It is seen that both sides give the same result in Maple. Thus, equation (1.6) is true.

Problems

1.9 Define each of the following vectors using Maple:

(a) $\mathbf{a} = (1, 1, -1)$

(b) $\mathbf{b} = (0, 2x, y^3)$

(c) $\mathbf{v} = (0, 0, 0, 0)$

(d) $\mathbf{w} = (y, y - 1)$

(e) A general three-dimensional vector $\mathbf{x} = (x_1, x_2, x_3)$

1.10 Consider the two vectors $\mathbf{u} = (2, 0, -1)$ and $\mathbf{v} = (3, 3, 3)$. Perform the following operations using Maple:

(a) $\mathbf{u} - 2$ (subtract 2 from each element of $\mathbf{u}$)

(b) $3\mathbf{v}$

(c) $\mathbf{u} + \mathbf{v}$

(d) $\mathbf{u} - \mathbf{v}$

(e) $3\mathbf{u} + 5\mathbf{v} - 1$

(f) $\mathbf{v} \cdot \mathbf{u}$

(g) $\mathbf{v} \times \mathbf{u}$

1.11 Consider the vector $\mathbf{u} = (1, -1, -1)$. Calculate the following norms using Maple:

(a) $\|\mathbf{u}\|_1$

(b) $\|\mathbf{u}\|_2$

(c) $\|\mathbf{u}\|_3$

(d) $\|\mathbf{u}\|_\infty$

1.12 Normalize the vector $\mathbf{u}=(8,-2,3)$ using Maple.

1.13 Determine the angle (in degrees) between the two vectors $\mathbf{a} = (2,0,1)$ and $\mathbf{b} = (1,-1,-2)$ using Maple.

1.14 Consider two two-dimensional vectors $\mathbf{u} = (u_1, u_2)$ and $\mathbf{v} = (v_1, v_2)$ as shown in Figure 1.5. Let θ be the angle between the two vectors. Prove equation (1.5) for these two vectors using the geometry of the problem.

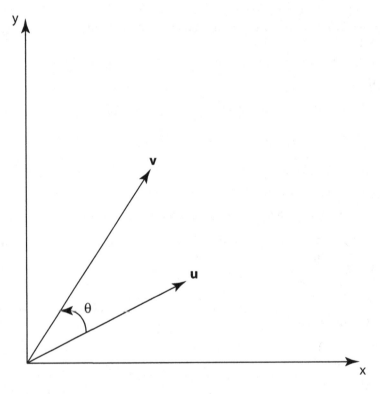

FIGURE 1.5

Two-dimensional vectors $\mathbf{u}$ and $\mathbf{v}$

1.15 Define two general three-dimensional vectors $\mathbf{u} = (u_1, u_2, u_3)$ and $\mathbf{v} = (v_1, v_2, v_3)$. Perform the following operations using Maple:

  (a) $\mathbf{u} + 2\mathbf{v}$

  (b) $\frac{3}{2}\mathbf{u}$

  (c) $(\mathbf{u} \times \mathbf{v}) + (\mathbf{v} \times \mathbf{u})$

  (d) Normalize $\mathbf{u} + \mathbf{v}$

  (e) Calculate the 2-norm of $\mathbf{u} - \mathbf{v}$

1.16 Consider two points $A = (5, 0, 2)$ and $B = (4, 4, 1)$ in three-dimensional space. Determine the vector **AB** starting at A and ending at B using Maple.

1.17 Determine the length of the vector $\mathbf{v} = (2, 0, 3)$ using Maple.

1.18 Consider two general three-dimensional vectors $\mathbf{u}$ and $\mathbf{v}$. Suppose that $\|\mathbf{u}\|_2 = \|\mathbf{v}\|_2$. Give an example where this holds but $\mathbf{u} \neq \mathbf{v}$.

1.19 Consider two general three-dimensional vectors $\mathbf{u}$ and $\mathbf{v}$. Use Maple to show that the angle θ between the two vectors can be calculated using the following formula:

$$\sin^2 \theta = \frac{(\mathbf{u} \cdot \mathbf{u})(\mathbf{v} \cdot \mathbf{v}) - (\mathbf{u} \cdot \mathbf{v})^2}{(\mathbf{u} \cdot \mathbf{u})(\mathbf{v} \cdot \mathbf{v})}$$

1.20 Use vectors and Maple to determine the angle between the face diagonals of a cube.

1.21 Consider three three-dimensional vectors $\mathbf{u}$, $\mathbf{v}$, and $\mathbf{w}$. Prove the following relationships using Maple.

  (a) $\mathbf{u} \cdot (\mathbf{v} \times \mathbf{w}) = (\mathbf{u} \times \mathbf{v}) \cdot \mathbf{w}$

  (b) $\mathbf{u} \times (\mathbf{v} \times \mathbf{w}) = \mathbf{v}(\mathbf{u} \cdot \mathbf{w}) - \mathbf{w}(\mathbf{u} \cdot \mathbf{v})$

  (c) $\mathbf{u} \times (\mathbf{v} \times \mathbf{w}) + \mathbf{v} \times (\mathbf{w} \times \mathbf{u}) + \mathbf{w} \times (\mathbf{u} \times \mathbf{v}) = \mathbf{0}$

  (d) $\mathbf{u} \times \mathbf{u} = \mathbf{0}$

  (e) $(\mathbf{u} + \mathbf{v}) \cdot (\mathbf{v} + \mathbf{w}) \times (\mathbf{w} + \mathbf{u}) = 2\mathbf{u} \cdot \mathbf{v} \times \mathbf{w}$

1.22 Consider four three-dimensional vectors $\mathbf{u}$, $\mathbf{v}$, $\mathbf{p}$, and $\mathbf{q}$. Prove the following relations using Maple:

  (a) $(\mathbf{u} \times \mathbf{v}) \times (\mathbf{p} \times \mathbf{q}) = \mathbf{p}(\mathbf{u} \times \mathbf{v} \cdot \mathbf{q}) - \mathbf{q}(\mathbf{u} \times \mathbf{v} \cdot \mathbf{p})$

  (b) $(\mathbf{u} \times \mathbf{v}) \cdot (\mathbf{p} \times \mathbf{q}) = (\mathbf{u} \cdot \mathbf{p})(\mathbf{v} \cdot \mathbf{q}) - (\mathbf{u} \cdot \mathbf{q})(\mathbf{v} \cdot \mathbf{p})$

1.23 Consider two vectors $\mathbf{u}$ and $\mathbf{v}$ satisfying the equation $\mathbf{u} = \alpha\mathbf{v}$ where α is a scalar. Show that

$$\alpha = \frac{\mathbf{u} \cdot \mathbf{v}}{|\mathbf{v}|^2} = \frac{|\mathbf{u}|^2}{\mathbf{u} \cdot \mathbf{v}}$$

1.24 Prove the *Law of Cosines* using vectors. (*Hint*: Consider the vector $\mathbf{w} = \mathbf{u} - \mathbf{v}$ and calculate the dot product $\mathbf{w} \cdot \mathbf{w}$).

1.25 Prove the *Law of Sines* using vectors. (*Hint*: Consider the cross products $\mathbf{u} \times \mathbf{v} = \mathbf{v} \times \mathbf{w} = \mathbf{w} \times \mathbf{u}$ and use $|\mathbf{u} \times \mathbf{v}| = |\mathbf{u}| \, |\mathbf{v}| \sin\theta$ where θ is the angle between $\mathbf{u}$ and $\mathbf{v}$, see Figure 1.6).

FIGURE 1.6
Three vectors $\mathbf{u}$, $\mathbf{v}$, and $\mathbf{w}$.

1.26 Consider three vectors $\mathbf{u}$, $\mathbf{v}$, and $\mathbf{w}$ such that $\mathbf{u}$ and $\mathbf{v}$ are orthogonal (i.e., $\mathbf{u} \cdot \mathbf{v} = 0$). Let θ be the angle between $\mathbf{u}$ and $\mathbf{v}$. Suppose the following relation holds:

$$\mathbf{w} = \alpha\mathbf{u} + \beta\mathbf{v}$$

where α and β are scalars. Show that α and β can be obtained using the following formulae:

$$\alpha = \frac{|\mathbf{w}|}{|\mathbf{u}|}\cos\theta$$

$$\beta = \frac{|\mathbf{w}|}{|\mathbf{v}|}\sin\theta$$

1.27 Consider a parallelogram with sides given by vectors $\mathbf{u}$ and $\mathbf{v}$. Show that the area of the parallelogram is given by $|\mathbf{u} \times \mathbf{v}| = |\mathbf{u}| \, |\mathbf{v}| \sin\theta$ where θ is the angle between $\mathbf{u}$ and $\mathbf{v}$.

1.28 Find the area of the triangle with vertices $(1, 2, -1)$, $(4, -5, 3)$, and $(0, 2, 1)$. Use Maple. (*Hint*: Use the vectors and the result of Problem 1.27 above.)

1.29 Consider a three-dimensional vector $\mathbf{u} = (x, y, z)$ which starts at the origin and ends at the point (x, y, z). Determine the angles α, β, and γ which the vector makes with the positive directions of the coordinate axes and show that:

$$\cos^2 \alpha + \cos^2 \beta + \cos^2 \gamma = 1$$

1.3 Matrices

In this section we explore matrices and matrix operations using Maple. In order to use matrices in Maple, you need first to invoke the linear algebra package as follows:

```
>  with(linalg);
```

Warning, the protected names norm and trace have been redefined and unprotected

[*BlockDiagonal*, *GramSchmidt*, *JordanBlock*, *LUdecomp*, *QRdecomp*, *Wronskian*, *addcol*, *addrow*, *adj*, *adjoint*, *angle*, *augment*, *backsub*, *band*, *basis*, *bezout*, *blockmatrix*, *charmat*, *charpoly*, *cholesky*, *col*, *coldim*, *colspace*, *colspan*, *companion*, *concat*, *cond*, *copyinto*, *crossprod*, *curl*, *definite*, *delcols*, *delrows*, *det*, *diag*, *diverge*, *dotprod*, *eigenvals*, *eigenvalues*, *eigenvectors*, *eigenvects*, *entermatrix*, *equal*, *exponential*, *extend*, *ffgausselim*, *fibonacci*, *forwardsub*, *frobenius*, *gausselim*, *gaussjord*, *geneqns*, *genmatrix*, *grad*, *hadamard*, *hermite*, *hessian*, *hilbert*, *htranspose*, *ihermite*, *indexfunc*, *innerprod*, *intbasis*, *inverse*, *ismith*, *issimilar*, *iszero*, *jacobian*, *jordan*, *kernel*, *laplacian*, *leastsqrs*, *linsolve*, *matadd*, *matrix*, *minor*, *minpoly*, *mulcol*, *mulrow*, *multiply*, *norm*, *normalize*, *nullspace*, *orthog*, *permanent*, *pivot*, *potential*, *randmatrix*, *randvector*, *rank*, *ratform*, *row*, *rowdim*, *rowspace*, *rowspan*, *rref*, *scalarmul*, *singularvals*, *smith*, *stackmatrix*, *submatrix*, *subvector*, *sumbasis*, *swapcol*, *swaprow*, *sylvester*, *toeplitz*, *trace*, *transpose*, *vandermonde*, *vecpotent*, *vectdim*, *vector*, *wronskian*]

Consider the following three matrices:

$$\mathbf{A} = \begin{bmatrix} 1 & 3 \\ -2 & 5 \end{bmatrix}$$

$$\mathbf{B} = \begin{bmatrix} 2 & x & y \\ x^2 & 2y & -1 \\ 0 & xy & x+y \end{bmatrix}$$

$$\mathbf{C} = \begin{bmatrix} 1 & r & r^2 & r^3 \\ 0 & 2 & -1 & 0 \\ r & r^2 & r^3 & r^4 \\ 1 & 0 & 2 & -r \end{bmatrix}$$

The above three matrices can be defined in Maple as follows:

```
> A:=matrix([[1,3],[-2,5]]);
```

$$A := \begin{bmatrix} 1 & 3 \\ -2 & 5 \end{bmatrix}$$

```
> B:=matrix([[2,x,y],[x^2,2*y,-1],[0,x*y,x+y]]);
```

$$B := \begin{bmatrix} 2 & x & y \\ x^2 & 2y & -1 \\ 0 & xy & x+y \end{bmatrix}$$

```
> C:=matrix([[1,r,r^2,r^3],[0,2,-1,0],[r,r^2,r^3,r^4],[1,0,2,-r]]);
```

$$C := \begin{bmatrix} 1 & r & r^2 & r^3 \\ 0 & 2 & -1 & 0 \\ r & r^2 & r^3 & r^4 \\ 1 & 0 & 2 & -r \end{bmatrix}$$

In addition to the Maple command *matrix* used above, the Maple command *array* can be used to define special types of matrices like the zero matrix and identity matrix as follows:

```
> A:=array(sparse,1..3,1..3);
```

$$A := \mathrm{array}(sparse,\ 1..3,\ 1..3,\ [])$$

```
> B:=array(identity,1..4,1..4);
```

$$B := \mathrm{array}(identity,\ 1..4,\ 1..4,\ [])$$

The Maple command *diag* can be used to define diagonal matrices as follows:

```
>   C:=diag(x,x^2,x^3);
```

$$C := \begin{bmatrix} x & 0 & 0 \\ 0 & x^2 & 0 \\ 0 & 0 & x^3 \end{bmatrix}$$

A banded matrix can be defined using the Maple command *band* as follows:

```
>   E:=band([1,2,-1],5);
```

$$E := \begin{bmatrix} 2 & -1 & 0 & 0 & 0 \\ 1 & 2 & -1 & 0 & 0 \\ 0 & 1 & 2 & -1 & 0 \\ 0 & 0 & 1 & 2 & -1 \\ 0 & 0 & 0 & 1 & 2 \end{bmatrix}$$

In the example above, the second argument indicates the size of the banded matrix.

The Jacobian matrix can be defined using the Maple command *jacobian* as follows:

```
>   f:=x^2+y^2+x^2;
```

$$f := 2\,x^2 + y^2$$

```
>   g:=x-y-z;
```

$$g := x - y - z$$

```
>   h:=3*x*y*z;
```

$$h := 3\,x\,y\,z$$

```
>   J:=jacobian([f,g,h],[x,y,z]);
```

$$J := \begin{bmatrix} 4x & 2y & 0 \\ 1 & -1 & -1 \\ 3\,y\,z & 3\,x\,z & 3\,x\,y \end{bmatrix}$$

In order to extract an element from a matrix, use the name of the matrix followed by square brackets. For example, the command $\mathbf{A}[1,2]$ will extract the element in row 1 and column 2 of the matrix $\mathbf{A}$. In order to extract a submatrix or a subvector, the Maple commands *submatrix* and *subvector* are used, respectively. These are illustrated by the following example:

```
>   B:=matrix([[3,5,7,9],[-1,0,1,5],[x,y,z,w],[r,p,q,s]]);
```

$$B := \begin{bmatrix} 3 & 5 & 7 & 9 \\ -1 & 0 & 1 & 5 \\ x & y & z & w \\ r & p & q & s \end{bmatrix}$$

> B[3,2];

$$y$$

> submatrix(B,2..3,3..4);

$$\begin{bmatrix} 1 & 5 \\ z & w \end{bmatrix}$$

> submatrix(B,[2,4],[1,3]);

$$\begin{bmatrix} -1 & 1 \\ r & q \end{bmatrix}$$

> subvector(B,3,1..4);

$$[x, y, z, w]$$

> subvector(B,[1,3,2],2);

$$[5, y, 0]$$

> row(B,3);

$$[x, y, z, w]$$

> col(B,4);

$$[9, 5, w, s]$$

The Maple command *row* and *col* were used above to extract the third row and the fourth column of matrix **B**, respectively.

The Maple command *transpose* is used to obtain the transpose of a matrix. The transpose of matrix **B** is obtained as follows:

> C:=transpose(B);

$$C := \begin{bmatrix} 3 & -1 & x & r \\ 5 & 0 & y & p \\ 7 & 1 & z & q \\ 9 & 5 & w & s \end{bmatrix}$$

The Maple commands *inverse* and *det* are used to obtain the inverse and determinant of a square matrix, respectively. The inverse and determinant of matrix **B** above are obtained as follows:

> B:=matrix([[3,6,3,8],[1,0,6,4],[1,2,3,4],[9,10,5,4]]);

$$B := \begin{bmatrix} 3 & 6 & 3 & 8 \\ 1 & 0 & 6 & 4 \\ 1 & 2 & 3 & 4 \\ 9 & 10 & 5 & 4 \end{bmatrix}$$

```
> C:=inverse(B);
```

$$C := \begin{bmatrix} 1 & 1 & \dfrac{-3}{} & 0 \\ \dfrac{-6}{7} & \dfrac{-13}{14} & \dfrac{71}{28} & \dfrac{3}{28} \\ \dfrac{-4}{7} & \dfrac{-2}{7} & \dfrac{19}{14} & \dfrac{1}{14} \\ \dfrac{17}{28} & \dfrac{3}{7} & \dfrac{-9}{7} & \dfrac{-3}{28} \end{bmatrix}$$

```
> d:=det(B);
```

$$d := 112$$

The trace (sum of the diagonal elements) of a square matrix can be obtained using the Maple command *trace*. The trace of matrix **B** above is calculated as follows:

```
> t:=trace(B);
```

$$t := 10$$

The Maple command *evalm* can be used to perform several matrix operations like matrix addition, matrix subtraction, scalar multiplication, and matrix multiplication, as shown in the following examples:

```
> A:=matrix([[2,5,8],[0,4,6],[1,2,4],[3,7,5]]);
```

$$A := \begin{bmatrix} 2 & 5 & 8 \\ 0 & 4 & 6 \\ 1 & 2 & 4 \\ 3 & 7 & 5 \end{bmatrix}$$

```
> B:=matrix([[1,2,6,7],[3,4,6,8],[1,1,1,5]]);
```

$$B := \begin{bmatrix} 1 & 2 & 6 & 7 \\ 3 & 4 & 6 & 8 \\ 1 & 1 & 1 & 5 \end{bmatrix}$$

```
> C:=evalm(A*2);
```

$$C := \begin{bmatrix} 4 & 10 & 16 \\ 0 & 8 & 12 \\ 2 & 4 & 8 \\ 6 & 14 & 10 \end{bmatrix}$$

```
>  E:=evalm(B-3);
```

$$E := \begin{bmatrix} -2\,2 & 6\,7 \\ 3\,1 & 6\,8 \\ 1\,1 & -2\,5 \end{bmatrix}$$

```
>  F:=evalm(A+transpose(B));
```

$$F := \begin{bmatrix} 3 & 8 & 9 \\ 2 & 8 & 7 \\ 7 & 8 & 5 \\ 10 & 15 & 10 \end{bmatrix}$$

```
>  G:=evalm(A&*B);
```

$$G := \begin{bmatrix} 25 & 32 & 50 & 94 \\ 18 & 22 & 30 & 62 \\ 11 & 14 & 22 & 43 \\ 29 & 39 & 65 & 102 \end{bmatrix}$$

```
>  H:=evalm(B&*A);
```

$$H := \begin{bmatrix} 29 & 74 & 79 \\ 36 & 99 & 112 \\ 18 & 46 & 43 \end{bmatrix}$$

```
>  J:=evalm(1/H);
```

$$J := \begin{bmatrix} \dfrac{895}{1277} & \dfrac{-452}{1277} & \dfrac{-467}{1277} \\ \dfrac{-468}{1277} & \dfrac{175}{1277} & \dfrac{404}{1277} \\ \dfrac{126}{1277} & \dfrac{2}{1277} & \dfrac{-207}{1277} \end{bmatrix}$$

It should be noted from the above examples that the operator & is used to indicate matrix multiplication. Multiplying a matrix and a vector is also easily performed in Maple as shown in the following examples:

```
>  A:=matrix([[7,4,5],[2,2,6],[-7,5,-5]]);
```

$$A := \begin{bmatrix} 7 & 4 & 5 \\ 2 & 2 & 6 \\ -7 & 5 & -5 \end{bmatrix}$$

```
>  e:=vector([2,-1,3]);
```

$$e := [2, -1, 3]$$

```
>  c:=evalm(A&*e);
```

$$c := [25, 20, -34]$$

```
>  d:=vector([[2],[-1],[3]]);
```

$$d := [[2], [-1], [3]]$$

> `f:=evalm(A&*d);`

$$f := \begin{bmatrix} 25 \\ 20 \\ -34 \end{bmatrix}$$

However, in order to multiply a vector and a matrix, the Maple command *innerprod* must be used. The following example uses **A** and **b** as defined above to multiply the vector **b** and the matrix **A**:

> `j:=innerprod(b,A);`

$$j := [-9, 21, -11]$$

Matrices can be used to solve systems of linear simultaneous equations. Consider the following system of equations:

$$2x - y + 3z = 4$$

$$x + 3y + 5z = -1$$

$$y - z = 2$$

The above system of equations can be written in matrix form as **Ax=b**, where **A**, **B**, and **x** are given by:

$$\mathbf{A} = \begin{bmatrix} 2 & -1 & 3 \\ 1 & 3 & 5 \\ 0 & 1 & -1 \end{bmatrix}$$

$$\mathbf{b} = \begin{bmatrix} 4 \\ -1 \\ 2 \end{bmatrix}$$

$$\mathbf{x} = \begin{bmatrix} x \\ y \\ z \end{bmatrix}$$

In order to solve the above system of linear equations, the Maple command *linsolve* is used as follows:

```
>  A:=matrix([[2,-1,3],[1,3,5],[0,1,-1]]);
```

$$A := \begin{bmatrix} 2 & -1 & 3 \\ 1 & 3 & 5 \\ 0 & 1 & -1 \end{bmatrix}$$

```
>  b:=vector([4,-1,2]);
```

$$b := [4, -1, 2]$$

```
>  x:=linsolve(A,b);
```

$$x := \begin{bmatrix} \dfrac{31}{7}, & \dfrac{4}{7}, & \dfrac{-10}{7} \end{bmatrix}$$

It should be noted that the command *linsolve* is used only to solve linear systems. Alternatively, one can use the inverse matrix of **A** to obtain the solution as follows:

```
>  x:=evalm(inverse(A)&*b);
```

$$x := \begin{bmatrix} \dfrac{31}{7}, & \dfrac{4}{7}, & \dfrac{-10}{7} \end{bmatrix}$$

It should be noted that using the inverse matrix takes more time to perform the calculations than using the command *linsolve*.

The eigenvalue of a matrix **A** are scalar values λ such that the equation

$$\mathbf{Ax} = \lambda \mathbf{x} \tag{1.7}$$

has one or more solution vectors **x**. The solution vectors **x** are called the eigenvectors of **A**. The eigenvalues and eigenvectors of a matrix are calculated using the Maple commands *eigenvals* and *eigenvects*, respectively. This is illustrated as follows:

```
>  A:=matrix([[1.2,-2.4,0],[2.3,3.6,1.1],[-2.0,3.0,1.4]]);
```

$$A := \begin{bmatrix} 1.2 & -2.4 & 0 \\ 2.3 & 3.6 & 1.1 \\ -2.0 & 3.0 & 1.4 \end{bmatrix}$$

```
>  lambda:=eigenvals(A);
```

$$\lambda := 1.234939238 + 1.588073742\,I,\ 1.234939238 - 1.588073742\,I,\ 3.730121523$$

```
>  x:=eigenvects(A);
```

$x := [1.234939240 + 1.588073741\,I, 1, \{[-.9558375413 - .8201580128\,I,$
$-.5287813216 + .6444150787\,I, 2.205134200 - .4340555658\,I]\}], [$
$1.234939240 - 1.588073741\,I, 1, \{[-.9558375413 + .8201580128\,I,$
$-.5287813216 - .6444150787\,I, 2.205134200 + .4340555658\,I]\}],$
$[3.730121526, 1, \{[-.5919850763, .6240809050, 1.311610933]\}]$

The orthonomal basis for a given set of vectors can be obtained using the Maple command *GramSchmidt* as follows:

```
>  v1:=vector([1,2,3,4,5]);
```
$$v1 := [1, 2, 3, 4, 5]$$
```
>  v2:=vector([-1,0,1,2,1]);
```
$$v2 := [-1, 0, 1, 2, 1]$$
```
>  v3:=vector([1,-1,1,-1,2]);
```
$$v3 := [1, -1, 1, -1, 2]$$
```
>  GramSchmidt([v1,v2,v3]);
```
$$[[1, 2, 3, 4, 5], [\frac{-14}{11}, \frac{-6}{11}, \frac{2}{11}, \frac{10}{11}, \frac{-4}{11}], [\frac{-1}{10}, \frac{-17}{10}, \frac{7}{10}, \frac{-9}{10}, 1]]$$

The resulting basis vectors above are normal to each other (use the Maple command *dotprod* to check this) but they are not unit vectors. Maple does not automatically normalize them to unit vectors. For this purpose, the *normalize* command should be used as was shown in Section 1.2.

General matrices can be defined in Maple even if their elements (or components) are not explicitly known. Consider the following examples and operations on general matrices using Maple:

```
>  A:=matrix(3,3);
```
$$A := \text{array}(1..3,\ 1..3,\ [])$$
```
>  B:=matrix(3,3);
```
$$B := \text{array}(1..3,\ 1..3,\ [])$$
```
>  x:=subvector(A,2,1..3);
```
$$x := [A_{2,1},\ A_{2,2},\ A_{2,3}]$$
```
>  C:=transpose(B);
```
$$C := \begin{bmatrix} B_{1,1} & B_{2,1} & B_{3,1} \\ B_{1,2} & B_{2,2} & B_{3,2} \\ B_{1,3} & B_{2,3} & B_{3,3} \end{bmatrix}$$
```
>  d:=det(B);
```

$$d := B_{1,1} B_{2,2} B_{3,3} - B_{1,1} B_{2,3} B_{3,2} - B_{2,1} B_{1,2} B_{3,3} + B_{2,1} B_{1,3} B_{3,2} + B_{3,1} B_{1,2} B_{2,3}$$
$$- B_{3,1} B_{1,3} B_{2,2}$$

```
>  t:=trace(A);
```

$$t := A_{1,1} + A_{2,2} + A_{3,3}$$

```
>  E:=evalm(A+2*B);
```

$$E := \begin{bmatrix} A_{1,1} + 2 B_{1,1} & A_{1,2} + 2 B_{1,2} & A_{1,3} + 2 B_{1,3} \\ A_{2,1} + 2 B_{2,1} & A_{2,2} + 2 B_{2,2} & A_{2,3} + 2 B_{2,3} \\ A_{3,1} + 2 B_{3,1} & A_{3,2} + 2 B_{3,2} & A_{3,3} + 2 B_{3,3} \end{bmatrix}$$

```
>  F:=evalm(A&*B);
```

$$F :=$$
$$\left[A_{1,1} B_{1,1} + A_{1,2} B_{2,1} + A_{1,3} B_{3,1},\ A_{1,1} B_{1,2} + A_{1,2} B_{2,2} + A_{1,3} B_{3,2},\right.$$
$$\left. A_{1,1} B_{1,3} + A_{1,2} B_{2,3} + A_{1,3} B_{3,3} \right]$$
$$\left[A_{2,1} B_{1,1} + A_{2,2} B_{2,1} + A_{2,3} B_{3,1},\ A_{2,1} B_{1,2} + A_{2,2} B_{2,2} + A_{2,3} B_{3,2},\right.$$
$$\left. A_{2,1} B_{1,3} + A_{2,2} B_{2,3} + A_{2,3} B_{3,3} \right]$$
$$\left[A_{3,1} B_{1,1} + A_{3,2} B_{2,1} + A_{3,3} B_{3,1},\ A_{3,1} B_{1,2} + A_{3,2} B_{2,2} + A_{3,3} B_{3,2},\right.$$
$$\left. A_{3,1} B_{1,3} + A_{3,2} B_{2,3} + A_{3,3} B_{3,3} \right]$$

Defining general matrices in Maple is very helpful especially in deriving complicated equations involving matrices.

Example 1.2

Let $\mathbf{M}$ be a matrix and $\mathbf{v}$ be a vector given by:

$$\mathbf{A} = \begin{bmatrix} 2 & -1 & 0 \\ -1 & 2 & 1 \\ 0 & -1 & 2 \end{bmatrix}$$

$$\mathbf{v} = \begin{bmatrix} 1 \\ x \\ x^2 \end{bmatrix}$$

Calculate the product $\mathbf{v^T M v}$ using Maple.

Solution

The following are the necessary Maple commands to calculate $\mathbf{v}^T\mathbf{Mv}$:

```
> M:=matrix([[2,-1,0],[-1,2,-2],[0,-1,2]]);
```

$$M := \begin{bmatrix} 2 & -1 & 0 \\ -1 & 2 & -2 \\ 0 & -1 & 2 \end{bmatrix}$$

```
> v:=vector([1,x,x^2]);
```

$$v := [1, \, x, \, x^2]$$

```
> c:=innerprod(v,M,v);
```

$$c := 2 - 2\,x + 2\,x^2 - 3\,x^3 + 2\,x^4$$

Example 1.3

(a) Consider a general square matrix $\mathbf{A}$. The adjoint matrix of $\mathbf{A}$, called adj$\mathbf{A}$, is defined by:

$$\mathbf{A}(\text{adj}\mathbf{A}) = (\text{adj}\mathbf{A})\mathbf{A} = |\,\mathbf{A}\,|\,\mathbf{I} \tag{1.8}$$

where $|\,\mathbf{A}\,|$ is the determinant of $\mathbf{A}$. Use the definition above to find a general expression for the adjoint matrix of $\mathbf{A}$.

(b) Find the adjoint matrix of $\mathbf{A}$ if $\mathbf{A}$ is given by:

$$\mathbf{A} = \begin{bmatrix} 1 & x & x^2 \\ 0 & -1 & 2 \\ 1 & 0 & x \end{bmatrix}$$

Solution

(a) Divide both sides of equation (1.8) by $|\mathbf{A}|$:

$$\frac{1}{|\,\mathbf{A}\,|}\mathbf{A}(\text{adj}\mathbf{A}) = \frac{1}{|\,\mathbf{A}\,|}(\text{adj}\mathbf{A})\mathbf{A} = \mathbf{I}$$

Rewrite the above equation as follows:

$$\mathbf{A}(\frac{\text{adj}\mathbf{A}}{|\,\mathbf{A}\,|}) = (\frac{\text{adj}\mathbf{A}}{|\,\mathbf{A}\,|})\mathbf{A} = \mathbf{I}$$

However, we know that

$$\mathbf{A}\mathbf{A}^{-1} = \mathbf{A}^{-1}\mathbf{A} = \mathbf{I}$$

Therefore, we conclude that

$$\mathbf{A}^{-1} = \frac{\mathrm{adj}\mathbf{A}}{|\,\mathbf{A}\,|}$$

The adjoint matrix of $\mathbf{A}$ is then given by:

$$\mathrm{adj}\mathbf{A} = |\,\mathbf{A}\,|\,\mathbf{A}^{-1} \tag{1.9}$$

(b) The following are the Maple commands to find the adjoint matrix of the given matrix $\mathbf{A}$ using equation (1.9).

```
>  A:=matrix([[1,x,x^2],[0,-1,2],[1,0,x]]);
```

$$A := \begin{bmatrix} 1 & x & x^2 \\ 0 & -1 & 2 \\ 1 & 0 & x \end{bmatrix}$$

```
>  adjoint_matrix:=det(A)*inverse(A);
```

$$adjoint\_matrix := (x + x^2) \begin{bmatrix} -\dfrac{1}{1+x} & -\dfrac{x}{1+x} & \dfrac{2+x}{1+x} \\[2mm] 2\,\dfrac{1}{x\,(1+x)} & -\dfrac{x-1}{1+x} & -2\,\dfrac{1}{x\,(1+x)} \\[2mm] \dfrac{1}{x\,(1+x)} & \dfrac{1}{1+x} & -\dfrac{1}{x\,(1+x)} \end{bmatrix}$$

Example 1.4

Consider the matrix $\mathbf{M}$ given by:

$$\mathbf{M} = \begin{bmatrix} m & -m & 0 \\ 1 & -1 & m \\ 1 & m & -m \end{bmatrix}$$

Determine the values of m for which the determinant of $\mathbf{M}$ vanishes, i.e., $|\,\mathbf{M}\,| = 0$.

Solution

The Maple commands *det* and *fsolve* are used to solve the problem as follows:

```
>  M:=matrix([[m,-m,0],[1,-1,m],[1,m,-m]]);
```

$$M := \begin{bmatrix} m & -m & 0 \\ 1 & -1 & m \\ 1 & m & -m \end{bmatrix}$$

```
>  y:=det(M);
```

$$y := -m^2 - m^3$$

```
>  fsolve(y,m);
```

$$-1., 0., 0.$$

Problems

1.30 Define the following matrices using Maple:

(a)

$$\mathbf{A} = \begin{bmatrix} 1 & r \\ r^2 & r^3 \end{bmatrix}$$

(b)

$$\mathbf{B} = \begin{bmatrix} 2 & 1 & 0 \\ -1 & 3 & 2 \\ -4 & 5 & 3 \end{bmatrix}$$

(c) an identity matrix $\mathbf{C}$ of size 6×6.

(d) a zero matrix $\mathbf{E}$ of size 2×2.

(e) a diagonal matrix $\mathbf{F}$ of size 4×4 where the diagonal elements are 1, 2, -1, 3.

1.31 Consider the matrix $\mathbf{R}$ given by

$$\mathbf{B} = \begin{bmatrix} 1 & z & -2 & y-x & \frac{x}{y} \\ 2 & 3 & 1 & y^3 & x \\ x & -1 & 1 & xy & z^2 \\ y & 2 & 0 & 2 & 1 \\ 0 & x^2 & 5 & 3 & 5 \end{bmatrix}$$

Extract the following components of $\mathbf{R}$ using Maple:

(a) the element in row 2 and column 4.

(b) the submatrix in rows 1 to 3 and columns 2 to 3.

(c) the submatrix in rows 2 and 5 and columns 4 and 5.

(d) the subvector in row 4 and columns 2, 4, 5, and 1 (exactly in the order given).

(e) the subvector in column 3.

(f) row 2.

(g) column 5.

1.32 Determine, using Maple, the transpose, inverse, determinant, and trace of matrix $\mathbf{X}$ given by

$$\mathbf{X} = \begin{bmatrix} 1 & x & -x \\ 2x & 2 & x \\ 1 & 0 & x \end{bmatrix}$$

1.33 Consider the two matrices $\mathbf{M}$ and $\mathbf{N}$ given by

$$\mathbf{M} = \begin{bmatrix} 1 & x \\ y & 0 \\ x & y \end{bmatrix}$$

$$\mathbf{N} = \begin{bmatrix} 1 & x & y \\ 0 & y & x \end{bmatrix}$$

Perform the following operations using Maple:

(a) $\mathbf{M} + x - y$ (add the scalar quantity x-y to each element of $\mathbf{M}$)

(b) $\mathbf{MN}$

(c) $\mathbf{N} + \mathbf{M}^{\mathrm{T}}$

(d) $\mathbf{M} + \mathbf{N}^{\mathrm{T}}$

(e) $(\mathbf{NM})^{-1}$

1.34 Solve the following system of equations using matrices and Maple. Use both methods of the inverse matrix and the Maple command *linsolve* independently.

$$3u + v - w = 0$$

$$u - v + 2w = 1$$

$$5u + 2v - 3w = -2$$

1.35 Determine using Maple the determinant and inverse of the following Jacobian matrix $\mathbf{J}$ where $f(x, y) = x^2 + y^2$ and $g(x, y) = \sin(xy)$.

$$\mathbf{J} = \begin{bmatrix} \dfrac{\partial f}{\partial x} & \dfrac{\partial f}{\partial y} \\ \dfrac{\partial g}{\partial x} & \dfrac{\partial g}{\partial y} \end{bmatrix}$$

1.36 Let $\mathbf{A}$ and $\mathbf{v}$ be defined as a matrix and a vector, respectively, as follows:

$$\mathbf{A} = \begin{bmatrix} x^2 & y^2 & 0 \\ 1 & xy & -1 \\ 1 & x & y \end{bmatrix}$$

$$\mathbf{v} = \begin{bmatrix} 1 \\ -1 \\ 0 \end{bmatrix}$$

Determine the three products $\mathbf{Av}$, $\mathbf{v}^T\mathbf{A}$, and $\mathbf{v}^T\mathbf{Av}$ using Maple.

1.37 Determine the eigenvalues and eigenvectors for each one of the following matrices using Maple:

(a)

$$S = \begin{bmatrix} 1 & 2 \\ 3 & 4 \end{bmatrix}$$

(b)

$$T = \begin{bmatrix} 2 & 0 & 1 \\ -3 & 4 & 2 \\ -1 & 0 & 5 \end{bmatrix}$$

1.38 Determine the orthonormal basis for the following set of vectors using Maple:

$$v_1 = [1, -2, 3]^T$$

$$v_2 = [0, 4, 5]^T$$

$$v_3 = [2, 2, 7]^T$$

1.39 Define two general square matrices $\mathbf{P}$ and $\mathbf{Q}$, each of size 2×2, with their elements unspecified, using Maple. Perform the following operations using Maple:

(a) $\mathbf{P} - \mathbf{Q}$
(b) $\mathbf{P}^{-1}$
(c) $(\mathbf{PQ})^T$
(d) adj$\mathbf{Q}$ (use the result of Example 1.3)
(e) $|\mathbf{P}^{-1}\mathbf{Q}|$

1.40 Consider two general matrices $\mathbf{A}$ and $\mathbf{B}$ of size 2×3. Show that $\alpha(\mathbf{A} + \mathbf{B}) = \alpha\mathbf{A} + \alpha\mathbf{B}$ using Maple, where α is a scalar.

1.41 Consider two general matrices $\mathbf{A}$ and $\mathbf{B}$ of sizes 2×3 and 3×4, respectively. Show that $(\mathbf{AB})^T = \mathbf{B}^T\mathbf{A}^T$ using Maple.

1.42 Let $\mathbf{A}$ be a 2 × 2 matrix given by

$$A = \begin{bmatrix} 1 & 0 \\ 2 & 3 \end{bmatrix}$$

Determine $\mathbf{A}^2$ and $\sqrt{\mathbf{A}}$ using Maple.

1.43 Consider a 2×2 general matrix $\mathbf{A}$. Show that $(\mathbf{A}^T)^{-1} = (\mathbf{A}^{-1})^T$ using Maple.

1.44 Three matrices $\mathbf{A}$, $\mathbf{B}$, and $\mathbf{C}$ are said to be linearly independent if the linear combination $\alpha \mathbf{A} + \beta \mathbf{B} + \gamma \mathbf{C} = 0$ implies that the undetermined scalar coefficients α, β, and γ vanish, i.e. $\alpha = \beta = \gamma = 0$. Are the following three matrices linearly independent? Use Maple.

$$\mathbf{A} = \begin{bmatrix} 1 & 1 \\ 1 & 1 \end{bmatrix}$$

$$\mathbf{B} = \begin{bmatrix} 1 & 0 \\ 0 & 1 \end{bmatrix}$$

$$\mathbf{C} = \begin{bmatrix} 0 & 0 \\ 1 & 1 \end{bmatrix}$$

1.45 A square matrix is said to be singular if its determinant vanishes. Give an example of a 3×3 singular matrix.

1.46 A matrix $\mathbf{Q}$ is said to be orthogonal if $\mathbf{Q}^T = \mathbf{Q}^{-1}$. Which of the following matrices is orthogonal? Use Maple.

(a)

$$\mathbf{Q} = \begin{bmatrix} \cos\theta & \sin\theta \\ -\sin\theta & \cos\theta \end{bmatrix}$$

(b)

$$\mathbf{Q} = \begin{bmatrix} \cos\theta & \sin\theta \\ \sin\theta & \cos\theta \end{bmatrix}$$

1.47 Two matrices $\mathbf{A}$ and $\mathbf{B}$ are related by the equation $\mathbf{A} = \mathbf{P}^{-1}\mathbf{B}\mathbf{P}$ where

$$\mathbf{A} = \begin{bmatrix} -30 & -48 \\ 18 & 29 \end{bmatrix}$$

$$B = \begin{bmatrix} 0 & 2 \\ 3 & -1 \end{bmatrix}$$

Determine the matrix **P** using Maple.

1.48 Given a diagonal 3×3 matrix **A** as follows:

$$A = \begin{bmatrix} a_1 & 0 & 0 \\ 0 & a_2 & 0 \\ 0 & 0 & a_3 \end{bmatrix}$$

Determine an expression for $\sqrt{A}$ using Maple.

1.49 Show that the determinant of the product of two 3×3 general matrices is equal to the product of their determinants, i.e., $| \mathbf{AB} | = | \mathbf{A} \| \mathbf{B} |$ for any two 3×3 general matrices **A** and **B**. Use Maple.

1.50 Let **A** be a square matrix of size $n \times n$ where n is a positive integer. Let $v_1, v_2, ..., v_n$ be the eigenvectors of **A**. Define the matrix **P** as follows:

$$\mathbf{P} = [\mathbf{v}_1, \mathbf{v}_2 ... \mathbf{v}_n]$$

P is defined above as the matrix of eigenvectors. Then the product $\mathbf{P}^{-1}\mathbf{AP}$ is a diagonal matrix. Diagonalize the matrix **A** given below using Maple:

$$A = \begin{bmatrix} 1 & 2 & 3 \\ -1 & 2 & -1 \\ 2 & 3 & 4 \end{bmatrix}$$

1.51 The characteristic polynomial of a matrix **A** is defined by the determinant $| \lambda \mathbf{I} - \mathbf{A} |$ where λ is a scalar. Determine the characteristic polynomial of **A** given below using Maple:

$$A = \begin{bmatrix} 1 & 4 & 0 \\ -2 & 3 & 5 \\ 1 & 2 & 5 \end{bmatrix}$$

1.52 Use the definition of the characteristic polynomial in Problem 1.51 to show that every general matrix of size 2×2 is a zero of its characteristic polynomial. Use Maple. (This is called the *Cayley-Hamilton Theorem* in Linear Algebra where it applies to any general matrix of size $n \times n$).

1.53 Use Maple and the definition of the characteristic polynomial in Problem 1.51 to show that the eigenvalues of a matrix are the roots of its characteristic polynomial. Consider a general 3×3 square matrix.

1.54 Let $\mathbf{A}$ be a 2×2 general square matrix as follows:

$$\mathbf{A} = \begin{bmatrix} a_{11} & a_{12} \\ a_{21} & a_{22} \end{bmatrix}$$

(a) Determine the two eigenvalues λ_1 and λ_2 of $\mathbf{A}$.

(b) Determine the inverse matrix $\mathbf{A}^{-1}$.

(c) Determine the two eigenvalues η_1 and η_2 of $\mathbf{A}^{-1}$.

(d) Show that $\eta_1 = \dfrac{1}{\lambda_1}$ and $\eta_2 = \dfrac{1}{\lambda_2}$

1.4 Indicial Notation

In this section we introduce the indicial notation to be used throughout the book. The purpose of using indicial notation is to write compact equations that can easily be handled.

Consider the variables $x_1, x_2, x_3, ..., x_n$ where n is a positive integer. The notation x_i can be used to denote these variables where the subscript i (called an index) is free to take the values 1, 2, 3,..., n. Consider now the notation a_{ij} where indices i and j are free to take the values from 1 to n. The notation a_{ij} is used to denote the $n \times n$ variables $a_{11}, a_{12}, ..., a_{1n}, a_{21}, a_{22}, ..., a_{2n}, ..., a_{n1}, a_{n2}, a_{nn}$. In this book we will always use $n = 3$. Thus the notation x_i will denote the three variables x_1, x_2, and x_3, while the notation a_{ij} will denote the nine variables $a_{11}, a_{12}, a_{13}, a_{21}, a_{22}, a_{23}, a_{31}, a_{32}$, and a_{33}.

Consider now the following equation written in indicial notation:

$$y_i = a_{ij}x_j \tag{1.10}$$

It is noted that in each term of the equation, the index i appears only once. Note also that the index j is repeated twice in the term on the right hand side of the equation. A repeated index indicates summation over this index in the range of 1 to 3. Thus equation (1.10) is equivalent to the following equation:

$$y_i = \Sigma_{j=1}^3 a_{ij} x_j \qquad (1.11)$$

According to the Einstein summation convention, the summation sign, Σ, is dropped from the equation. Therefore, the notation of equation (1.10) will be used throughout the book. The following are the rules for the Einstein summation convention and indicial notation:

1. A repeated index indicates summation over this index in the range of 1, 2, and 3. The repeated index is called a *dummy* index because the letter used for this index can be replaced by any other letter. In this case, the summation sign, Σ, is dropped from the equation.

2. An index that appears only once in every term of the equation is called a *free* index. Each free index used indicates three separate equations when the free index takes each of the values 1, 2, and 3.

3. No index can be repeated more than twice using this indicial notation. If there is a summation over an index that is repeated three or more times, then the summation sign, Σ, must be used.

We will now consider some examples illustrating the above three rules of indicial notation. According to Rule 1, equation (1.10) is equivalent to the following equation where the dummy index j is replaced first by m then by n:

$$y_i = a_{im} x_m = a_{in} x_n \qquad (1.12)$$

The validity of equation (1.12) is made clear by retaining the summation sign and rewriting the equation as follows:

$$y_i = \Sigma_{m=1}^3 a_{im} x_m = \Sigma_{n=1}^3 a_{in} x_n \qquad (1.13)$$

Next, we will write equation (1.10) explicitly in its expanded form as follows:

$$y_i = a_{i1} x_1 + a_{i2} x_2 + a_{i3} x_3 \qquad (1.14)$$

According to Rule 2 of the indicial notation, equation (1.10) actually represents three separate equations when the free index i takes each of the three values 1, 2, and 3. The three equations represented by equation (1.10) are:

$$y_1 = a_{1j}x_j \tag{1.15a}$$

$$y_2 = a_{2j}x_j \tag{1.15b}$$

$$y_3 = a_{3j}x_j \tag{1.15c}$$

Next, we expand equations (1.15) further by summing over the dummy index (or use equation (1.14)) to obtain:

$$y_1 = a_{11}x_1 + a_{12}x_2 + a_{13}x_3 \tag{1.16a}$$

$$y_2 = a_{21}x_1 + a_{22}x_2 + a_{23}x_3 \tag{1.16b}$$

$$y_1 = a_{31}x_1 + a_{32}x_2 + a_{33}x_3 \tag{1.16c}$$

It should be noted that equations (1.16) are the expanded form of equation (1.10).

According to Rule 3 of the indicial notation, the term $x_i y_i z_i$ has no significance and should not be used because the index i is repeated three times. In this case, the summation sign, Σ, should be retained and the term should be written as $\Sigma_{i=1}^{3} x_i y_i z_i$. The same rule applies to any index repeated more than twice. Consider now the following equation:

$$w = \lambda_{ij} u_i u_j \tag{1.17}$$

Equation (1.17) represents one equation only because there are no free indices in the terms of the equation. However, the term on the right side includes two dummy indices, i and j (each is repeated twice). Therefore, the right hand side can be expanded as follows. First, we expand the sum over i:

$$w = \lambda_{1j} u_1 u_j + \lambda_{2j} u_2 u_j + \lambda_{3j} u_3 u_j \tag{1.18}$$

Next, we expand each term on the right hand side of the equation (1.18) over j to obtain:

$$\begin{aligned} w = \lambda_{11} u_1 u_1 + \lambda_{12} u_1 u_2 + \lambda_{13} u_1 u_3 \\ + \lambda_{21} u_2 u_1 + \lambda_{22} u_2 u_2 + \lambda_{23} u_2 u_3 \\ + \lambda_{31} u_3 u_1 + \lambda_{32} u_3 u_2 + \lambda_{33} u_3 u_3 \end{aligned} \tag{1.19}$$

It should be noted that equation (1.17) is exactly equivalent to the longer equation (1.19). In fact, equation (1.19) is the expanded form of equation (1.17).

The Kronecker delta, δ_{ij}, is defined by:

$$\delta_{ij} = \begin{cases} 1, i = j \\ 0, i \neq j \end{cases} \tag{1.20}$$

For example, $\delta_{11} = \delta_{22} = \delta_{33} = 1$, and $\delta_{12} = \delta_{23} = \delta_{31} = 0$. Therefore, δ_{ij} can be represented by the 3×3 identity matrix as follows:

$$\delta_{ij} \equiv \begin{bmatrix} 1 & 0 & 0 \\ 0 & 1 & 0 \\ 0 & 0 & 1 \end{bmatrix} \tag{1.21}$$

Note that $\delta_{ii} = \delta_{11} + \delta_{22} + \delta_{33} = 1 + 1 + 1 = 3$.

Next, we will explore the term $\delta_{ij}a_j$. The index i is a free index; therefore the term $\delta_{ij}a_j$ represents three terms, namely $\delta_{1j}a_j$, $\delta_{2j}a_j$, and $\delta_{3j}a_j$. Evaluating each term separately, we obtain:

$$\begin{aligned} \delta_{1j}a_j &= \delta_{11}a_1 + \delta_{12}a_2 + \delta_{13}a_3 \\ &= (1)a_1 + (0)a_2 + (0)a_3 \\ &= a_1 \end{aligned} \tag{1.22a}$$

$$\begin{aligned} \delta_{2j}a_j &= \delta_{21}a_1 + \delta_{22}a_2 + \delta_{23}a_3 \\ &= (0)a_1 + (1)a_2 + (0)a_3 \\ &= a_2 \end{aligned} \tag{1.22b}$$

$$\begin{aligned} \delta_{3j}a_j &= \delta_{31}a_1 + \delta_{32}a_2 + \delta_{33}a_3 \\ &= (0)a_1 + (0)a_2 + (1)a_3 \\ &= a_3 \end{aligned} \tag{1.22c}$$

It is now clear from equation (1.22) that the following relation is true:

$$\delta_{ij}a_j = a_i \tag{1.23}$$

Similarly, one can write nine equations to show that:

$$\delta_{im}A_{mj} = A_{ij} \tag{1.24}$$

The permutation symbol, ϵ_{ijk}, has one of the values 0, 1, or -1 depending on the permutation formed by i, j, and k. The following are the three rules for the permutation symbol ϵ_{ijk}:

1. If i, j, and k form an even permutation of 1, 2, and 3, then $\epsilon_{ijk} = 1$. For example, $\epsilon_{123} = \epsilon_{231} = \epsilon_{312} = 1$.

2. If i, j, and k form an odd permutation of 1, 2, and 3, then $\epsilon_{ijk} = -1$.
   For example, $\epsilon_{132} = \epsilon_{321} = \epsilon_{213} = -1$.

3. If i, j, and k do not form a permutation of 1, 2, and 3, then $\epsilon_{ijk} = 0$.
   For example, $\epsilon_{111} = \epsilon_{122} = \epsilon_{223} = 0$.

Using the three rules above, it is clear that:

$$\epsilon_{ijk} = \epsilon_{jki} = \epsilon_{kij} = -\epsilon_{jik} = -\epsilon_{kji} = -\epsilon_{ikj} \tag{1.25}$$

Example 1.5

Show that equation (1.24) is true.

Solution

Equation (1.24) has two free indices, namely, i and j. Therefore, this equation represents nine equations as follows:

$$\delta_{1m}A_{m1} = \delta_{11}A_{11} + \delta_{12}A_{21} + \delta_{13}A_{31} = A_{11}$$
$$\delta_{1m}A_{m2} = \delta_{11}A_{12} + \delta_{12}A_{22} + \delta_{13}A_{32} = A_{12}$$
$$\delta_{1m}A_{m3} = \delta_{11}A_{13} + \delta_{12}A_{23} + \delta_{13}A_{33} = A_{13}$$
$$\delta_{2m}A_{m1} = \delta_{21}A_{11} + \delta_{22}A_{21} + \delta_{23}A_{31} = A_{21}$$
$$\delta_{2m}A_{m2} = \delta_{21}A_{12} + \delta_{22}A_{22} + \delta_{23}A_{32} = A_{22}$$
$$\delta_{2m}A_{m3} = \delta_{21}A_{13} + \delta_{22}A_{23} + \delta_{23}A_{33} = A_{23}$$
$$\delta_{3m}A_{m1} = \delta_{31}A_{11} + \delta_{32}A_{21} + \delta_{33}A_{31} = A_{31}$$
$$\delta_{3m}A_{m2} = \delta_{31}A_{12} + \delta_{32}A_{22} + \delta_{33}A_{32} = A_{32}$$
$$\delta_{3m}A_{m3} = \delta_{31}A_{13} + \delta_{32}A_{23} + \delta_{33}A_{33} = A_{33} \tag{1.26}$$

It is clear from the equation above that $\delta_{im}A_{mj} = A_{ij}$

Example 1.6

Given $A_{ij} = -A_{ji}$, show that $A_{ij}v_iv_j = 0$.

Solution

We will start with $2A_{ij}v_iv_j$ as follows:

$$\begin{aligned}
2A_{ij}v_iv_j &= A_{ij}v_iv_j + A_{ij}v_iv_j \\
&= A_{ij}v_iv_j + A_{ji}v_jv_i \\
&= A_{ij}v_iv_j + (-A_{ij})v_jv_i \\
&= A_{ij}v_iv_j - A_{ij}v_iv_j \\
&= 0 \tag{1.27}
\end{aligned}$$

Therefore, $A_{ij} v_i v_j = 0$

In the second line of equation (1.27) above, we interchanged indices i and j in the second term because they are both dummy indices. Also, in the fourth line of the equation we have used the commutative property of scalars $v_j v_i = v_i v_j$.

Example 1.7

Given $A_{ij} = -A_{ji}$ and $B_{ij} = B_{ji}$, show that $A_{ij} B_{ij} = 0$.

Solution

We will start with $2 A_{ij} B_{ij}$ as follows:

$$
\begin{aligned}
2 A_{ij} B_{ij} &= A_{ij} B_{ij} + A_{ij} B_{ij} \\
&= A_{ij} B_{ij} + A_{ji} B_{ji} \\
&= A_{ij} B_{ij} + (-A_{ij}) B_{ij} \\
&= A_{ij} B_{ij} - A_{ij} B_{ij} \\
&= 0
\end{aligned} \tag{1.28}
$$

Therefore, $A_{ij} B_{ij} = 0$.

In the second line of equation (1.28) above, we interchanged indices i and j in the second term because they are both dummy indices.

Example 1.8

Let d be the determinant of a general 3×3 matrix $\mathbf{A} \equiv A_{ij}$. Show that

$$
d = \epsilon_{ijk} A_{i1} A_{j2} A_{k3}
$$

Solution

Let $\mathbf{A}$ be a 3×3 matrix given by:

$$
\mathbf{A} = \begin{bmatrix} A_{11} & A_{12} & A_{13} \\ A_{21} & A_{22} & A_{23} \\ A_{31} & A_{32} & A_{33} \end{bmatrix}
$$

Then, the determinant d of $\mathbf{A}$ can be written as follows:

$$
d = |\mathbf{A}| = \begin{vmatrix} A_{11} & A_{12} & A_{13} \\ A_{21} & A_{22} & A_{23} \\ A_{31} & A_{32} & A_{33} \end{vmatrix}
$$

$$= A_{11}(A_{22}A_{33} - A_{23}A_{32}) - A_{12}(A_{21}A_{33} - A_{23}A_{31}) + A_{13}(A_{21}A_{23} - A_{22}A_{31})$$
$$= A_{11}A_{22}A_{33} + A_{12}A_{23}A_{31} + A_{13}A_{21}A_{32} - A_{11}A_{23}A_{32} - A_{12}A_{21}A_{33} - A_{13}A_{31}A_{22}$$
$$= \epsilon_{123}A_{11}A_{22}A_{33} + \epsilon_{312}A_{31}A_{12}A_{23} + \epsilon_{231}A_{21}A_{32}A_{13}$$
$$+ \epsilon_{132}A_{11}A_{32}A_{23} + \epsilon_{213}A_{21}A_{12}A_{33} + \epsilon_{321}A_{31}A_{22}A_{13}$$
$$= \epsilon_{ijk}A_{i1}A_{j2}A_{k3} \tag{1.29}$$

Problems

1.55 Given that A_{ij} is represented by the following matrix:

$$A_{ij} \equiv \begin{bmatrix} 2 & -3 & 0 \\ 4 & 4 & 1 \\ -2 & 2 & 5 \end{bmatrix}$$

Determine the following quantities:

(a) A_{ii}

(b) $A_{ij}A_{ij}$

(c) $A_{ij}A_{ji}$

(d) $\delta_{ii}A_{mm}$

(e) $A_{pq}A_{pq}$

1.56 Write explicitly the three equations represented by the following equation:

$$v_m = Q_{mn}u_n$$

1.57 Write explicitly the nine equations represented by the equation:

$$A_{ij} = B_{ir}C_{rj}$$

1.58 How many equations are represented by the following equation. Write the expanded result explicitly.

$$S = v_i Q_{ij} v_j$$

1.59 Write the following three equations as one single equation using indicial notation:

$$dy_1 = \frac{\partial y_1}{\partial x_1} dx_1 + \frac{\partial y_1}{\partial x_2} dx_2 + \frac{\partial y_1}{\partial x_3} dx_3$$

$$dy_2 = \frac{\partial y_2}{\partial x_1} dx_1 + \frac{\partial y_2}{\partial x_2} dx_2 + \frac{\partial y_2}{\partial x_3} dx_3$$

$$dy_3 = \frac{\partial y_3}{\partial x_1} dx_1 + \frac{\partial y_3}{\partial x_2} dx_2 + \frac{\partial y_3}{\partial x_3} dx_3$$

1.60 Given that u_i, v_i, and A_{ij} are represented by the following two vectors and matrix:

$$u_i \equiv \begin{bmatrix} 1 \\ -1 \\ 0 \end{bmatrix}$$

$$v_i \equiv \begin{bmatrix} 2 \\ 3 \\ -5 \end{bmatrix}$$

$$A_{ij} \equiv \begin{bmatrix} 4 & 2 & -1 \\ -3 & 3 & 0 \\ 5 & 1 & 2 \end{bmatrix}$$

Determine the vectors or matrices represented by the following quantities:

(a) $w_k = \epsilon_{ijk} u_i v_j$

(b) $y_i = \epsilon_{ijk} A_{jk}$

(c) $B_{ij} = \epsilon_{ijk} u_k + \epsilon_{ijl} v_l$

(d) the scalar quantity $A_{ij} u_i v_j$

1.61 Given that $B_{ij} = \alpha A_{ij} + \beta \delta_{ij}$, where α and β are scalars. Show that:

(a) $B_{ij}A_{ij} = \alpha A_{ij}A_{ij} + \beta A_{mm}$

(b) $B_{ij}B_{ij} = \alpha^2 A_{ij}A_{ij} + 2\alpha\beta A_{mm} + 3\beta^2$

1.62 Show that the following identity holds between the Kronecker delta and the permutation symbol:

$$\epsilon_{ijk}\epsilon_{ipq} = \delta_{jp}\delta_{kq} - \delta_{jq}\delta_{kp}$$

1.63 Show that the following relations hold:

(a) $\epsilon_{ipq}\epsilon_{jpq} = 2\delta_{ij}$

(b) $\epsilon_{ijk}\epsilon_{ijk} = 6$ (*Hint*: Use the result of Problem 1.62 above).

1.64 Prove the following relations:

(a) $\delta_{mn}\delta_{mn} = 3$

(b) $\epsilon_{ijk}u_j u_k = 0$

(c) $\delta_{im}\delta_{mj} = \delta_{ij}$

(d) $\epsilon_{ijk}\delta_{ij} = 0$

1.65 Prove the following relation involving determinants:

$$\begin{vmatrix} A_{ip} & A_{iq} & A_{ir} \\ A_{jp} & A_{jq} & A_{jr} \\ A_{kp} & A_{kq} & A_{kr} \end{vmatrix} = \epsilon_{ijk}\epsilon_{pqr}|\mathbf{A}|$$

where $|\mathbf{A}|$ is the determinant of a general 3×3 matrix $\mathbf{A}$. Use this result to solve Problem 1.62 above directly.

1.66 Let $\mathbf{A}$ be a general 3×3 matrix whose elements are given by A_{ij}. The *minor* M_{ij} of an element A_{ij} is defined as the determinant of the matrix formed by deleting the ith row and the jth column of $\mathbf{A}$. The corresponding *cofactor* C_{ij} is defined by $C_{ij} = (-1)^{i+j}M_{ij}$. Show that the following two relations hold:

(a) $A_{pj}C_{ij} = |\mathbf{A}|\,\delta_{\mathrm{pi}}$

(b) $A_{ip}C_{ij} = |\mathbf{A}|\,\delta_{\mathrm{pj}}$

1.5   Transformation of Vectors

Let $\mathbf{e_1}$, $\mathbf{e_2}$, and $\mathbf{e_3}$ be three unit vectors along the x, y, and z coordinates, respectively (see Figure 1.7). Obviously, these unit vectors are perpendicular to each other so that $\mathbf{e_1} \cdot \mathbf{e_1} = 1$, $\mathbf{e_1} \cdot \mathbf{e_2} = 0$, etc. These relations are written using indicial notation as follows:

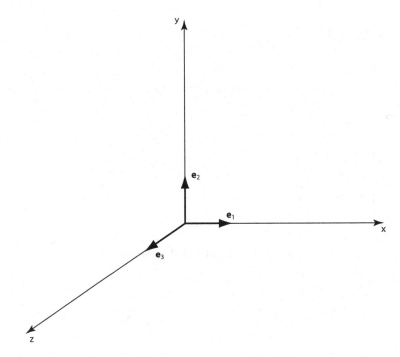

FIGURE 1.7

Unit vectors $\mathbf{e_1}$, $\mathbf{e_2}$, and $\mathbf{e_3}$.

$$\mathbf{e}_i \cdot \mathbf{e}_j = \delta_{ij} \tag{1.30}$$

Let $\mathbf{u}$ and $\mathbf{v}$ be two vectors such that

$$\mathbf{u} = u_1 \mathbf{e_1} + u_2 \mathbf{e_2} + u_3 \mathbf{e_3} \tag{1.31a}$$

$$\mathbf{v} = v_1 \mathbf{e_1} + v_2 \mathbf{e_2} + v_3 \mathbf{e_3} \tag{1.31b}$$

Equations (1.31) can be written using indicial notation as follows:

$$\mathbf{u} = u_i \mathbf{e_i} \tag{1.32a}$$

$$\mathbf{v} = v_i \mathbf{e_j} \tag{1.32b}$$

Next, consider the dot product $\mathbf{u} \cdot \mathbf{v}$. Using equations (1.30) and (1.32), we obtain:

$$\mathbf{u} \cdot \mathbf{v} = u_i v_j \delta_{ij} = u_i v_i \tag{1.33}$$

The three unit vectors have the property that $\mathbf{e_1} \times \mathbf{e_2} = \mathbf{e_3}$, $\mathbf{e_2} \times \mathbf{e_3} = \mathbf{e_1}$, and $\mathbf{e_1} \times \mathbf{e_1} = 0$, etc. This property can be written using indicial notation as follows:

$$\mathbf{e_i} \times \mathbf{e_j} = \epsilon_{ijk} \mathbf{e_k} \tag{1.34}$$

Consider now the cross product $\mathbf{u} \times \mathbf{v}$. Using equations (1.32) and (1.34), we obtain:

$$\mathbf{u} \times \mathbf{v} = \epsilon_{ijk} u_i v_j \mathbf{e_k} \tag{1.35}$$

Next, consider a *tranformation* $\mathbf{T}$ which transforms the vector $\mathbf{u}$ into the vector $\mathbf{v}$. This can be written as follows:

$$\mathbf{T}(\mathbf{u}) = \mathbf{v} \tag{1.36}$$

In general, the transformation $\mathbf{T}$ can be represented by a 3×3 matrix. For example, if $\mathbf{u} = \mathbf{v}$, then $\mathbf{T}$ would be represented by the identity matrix $\mathbf{I}$.

Let $\mathbf{v_1}$ and $\mathbf{v_2}$ be two vectors in three-dimensional space. Suppose that a transformation $\mathbf{T}$, when applied to $\mathbf{v_1}$ and $\mathbf{v_2}$, satisfies the following two properties of vector addition and scalar multiplication:

$$\mathbf{T}(\mathbf{v_1} + \mathbf{v_2}) = \mathbf{T}(\mathbf{v_1}) + \mathbf{T}(\mathbf{v_2}) \tag{1.37a}$$

$$\mathbf{T}(\alpha \mathbf{v_1}) = \alpha \mathbf{T}(\mathbf{v_1}) \tag{1.37b}$$

where α is a scalar. In this case, $\mathbf{T}$ is called a *linear transformation*.

Next, we will explore how to determine the components of vectors and linear transformations. Suppose that $\mathbf{u}$ and $\mathbf{v}$ are two vectors related by the linear transformation $\mathbf{T}$ as before, i.e., $\mathbf{v} = \mathbf{T}(\mathbf{u})$. Taking the dot product of the vector $\mathbf{u}$ with any of the unit vectors $\mathbf{e_1}$, $\mathbf{e_2}$, or $\mathbf{e_3}$ will result in the component of the vector along the respective coordinate axis. For example, $\mathbf{u} \cdot \mathbf{e_1} = u_1$, $\mathbf{u} \cdot \mathbf{e_2} = u_2$, and $\mathbf{u} \cdot \mathbf{e_3} = u_3$. These equations can be written using indicial notation as follows (see Figure 1.8):

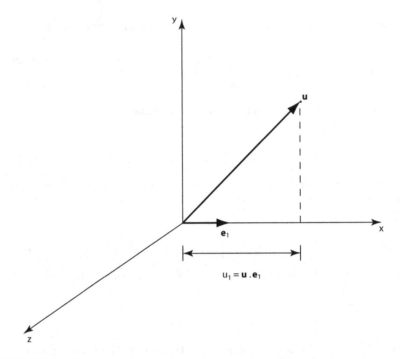

FIGURE 1.8
Projection of vector **u** on the x-axis.

$$u_i = \mathbf{u} \cdot \mathbf{e_i} \tag{1.38}$$

The components of the vector **u** can be obtained using equation (1.38) above. Next, we determine the components of the vector **v** and the linear transformation **T**.

Substituting equations (1.32a) into equations (1.36), we obtain:

$$
\begin{aligned}
\mathbf{u} = \mathbf{T}(\mathbf{u}) &= \mathbf{T}(u_i \mathbf{e_i} \\
&= \mathbf{T}(u_1 \mathbf{e_1} + u_2 \mathbf{e_2} + u_3 \mathbf{e_3}) \\
&= u_1 \mathbf{T}(\mathbf{e_1}) + u_2 \mathbf{T}(\mathbf{e_2}) + u_3 \mathbf{T}(\mathbf{e_3})
\end{aligned}
\tag{1.39}
$$

In deriving the last line of equation (1.39) above, we have used the fact that **T** is a linear transformation and applied both equations (1.37) simultaneously, where u_1, u_2, and u_3 are scalars.

Equation (1.39) can be written using indicial notation as follows:

$$\mathbf{v} = u_j \mathbf{T}(\mathbf{e_j}) \tag{1.40}$$

The components of vector $\mathbf{v}$ are written now using the dot product as shown in equation (1.38) but replacing $\mathbf{u}$ with $\mathbf{v}$.

$$v_i = \mathbf{v} \cdot \mathbf{e_i} = u_j \mathbf{T}(e_j) \cdot \mathbf{e_i} \tag{1.41}$$

In deriving equation (1.41), we have used equations (1.40) and the commutative property of the dot product. The dot product $\mathbf{T}(e_j) \cdot \mathbf{e_i}$ is the component of the vector $\mathbf{T}(e_j)$ along the direction of $\mathbf{e_i}$. We will denote this component by T_{ij}. Therefore, we have the following equation (by definition):

$$T_{ij} = \mathbf{T}(\mathbf{e_j}) \cdot \mathbf{e_i} \tag{1.42}$$

The components of the linear transformation $\mathbf{T}$ are obtained using equation (1.42) above. It is clear that $\mathbf{T}$ has nine components since equation (1.42) has two free indices and thus represents nine equations.

Finally, the components of vector $\mathbf{v}$ are obtained by substituting equation (1.42) into equation (1.41):

$$v_i = T_{ij} u_j \tag{1.43}$$

The linear transformation $\mathbf{T}$ can be represented by a 3×3 matrix since equation (1.43) can be represented in matrix form as follows:

$$\begin{bmatrix} v_1 \\ v_2 \\ v_3 \end{bmatrix} = \begin{bmatrix} T_{11} & T_{12} & T_{13} \\ T_{21} & T_{22} & T_{23} \\ T_{31} & T_{32} & T_{33} \end{bmatrix} \begin{bmatrix} u_1 \\ u_2 \\ u_3 \end{bmatrix} \tag{1.44}$$

It is clear from equations (1.42) and (1.44) that T_{11}, T_{21}, and T_{31} are the components of $\mathbf{T}(\mathbf{e_1})$. Therefore, the vector $\mathbf{T}(\mathbf{e_1})$ can be written as follows:

$$\mathbf{T}(\mathbf{e_1}) = T_{j1} \mathbf{e_j} \tag{1.45}$$

The same relation holds for the vectors $\mathbf{T}(\mathbf{e_2})$ and $\mathbf{T}(\mathbf{e_3})$. Therefore, equation (1.45) is generalized as follows:

$$\mathbf{T}(\mathbf{e_i}) = T_{ji} \mathbf{e_j} \tag{1.46}$$

Example 1.9

A linear transformation $\mathbf{T}$ transforms every vector $\mathbf{u}$ in three-dimensional space into a vector $\mathbf{v}$ such that $\mathbf{v} = \dfrac{1}{3}\mathbf{u}$. Determine the matrix represented by $\mathbf{T}$.

Solution

We have the following relations:

$$T(e_1) = (\frac{1}{3})e_1 + (0)e_2 + (0)e_3$$

$$T(e_2) = (0)e_1 + (\frac{1}{3})e_2 + (0)e_3$$

$$T(e_3) = (0)e_1 + (0)e_2 + (\frac{1}{3})e_3$$

Using equations (1.46), we obtain the matrix representation of T as follows:

$$T \equiv \begin{bmatrix} \frac{1}{3} & 0 & 0 \\ 0 & \frac{1}{3} & 0 \\ 0 & 0 & \frac{1}{3} \end{bmatrix} = \frac{1}{3}I$$

Example 1.10

Let T be a constant transformation that transforms every vector in three-dimensional space into the vector $e_1 + e_2 + e_3$. Is T a linear transformation? If yes, find the matrix representation of T.

Solution

Let u and v be two arbitrary vectors in three dimensional space. Then, we have the following relations:

$$T(u) = e_1 + e_2 + e_3$$

$$T(v) = e_1 + e_2 + e_3$$

We need to check if equations (1.37) hold.

$$T(u + v) = e_1 + e_2 + e_3$$

$$T(u) + T(v) = 2(e_1 + e_2 + e_3)$$

Thus it is clear that $T(u + v) \neq T(u) + T(v)$. Therefore, T is not a linear transformation.

Problems

1.67 Given $\mathbf{a} = 2\mathbf{e_1} - \mathbf{e_2} + 2\mathbf{e_3}$ and $\mathbf{b} = \mathbf{e} + 4\mathbf{e_3}$, determine the products $\mathbf{a} \cdot \mathbf{b}$ and $\mathbf{a} \times \mathbf{b}$ using equations (1.33) and (1.35), respectively.

1.68 Given the vector $\mathbf{u} = \mathbf{e_1} - 3\mathbf{e_2} + 5\mathbf{e_3}$ and the linear transformation $\mathbf{T}$ represented by the matrix:

$$\mathbf{T} \equiv \begin{bmatrix} 1 & -2 & 0 \\ -3 & 5 & 2 \\ 4 & 3 & 1 \end{bmatrix}$$

Determine the vector $\mathbf{v} = \mathbf{T(u)}$.

1.69 Consider the linear transformation $\mathbf{T}$ represented by the matrix

$$\mathbf{T} \equiv \begin{bmatrix} -3 & 1 & 0 \\ 4 & -2 & 3 \\ 1 & 4 & 5 \end{bmatrix}$$

What are the three components of the vector $\mathbf{T(e_2)}$.

1.70 Let $\mathbf{T}$ be a linear transformation that transforms every vector $\mathbf{u}$ into a vector $\mathbf{v}$ given by $\mathbf{v} = (\mathbf{u} \cdot \mathbf{a})\mathbf{b}$ where $\mathbf{a}$ and $\mathbf{b}$ are given by:

$\mathbf{a} = \mathbf{e_1} + \mathbf{e_2} + \mathbf{e_3}$

$\mathbf{b} = 2\mathbf{e_1} - \mathbf{e_2} + \mathbf{e_3}$

Determine the matrix representation of $\mathbf{T}$.

1.71 Let $\mathbf{T}$ be a transformation defined by $\mathbf{T(u)} = \alpha\mathbf{u} - \mathbf{e_1}$ where α is a scalar. Is $\mathbf{T}$ a linear transformation? If yes, determine the matrix representation of $\mathbf{T}$.

1.72 A linear transformation $\mathbf{T}$ is represented by the following matrix:

$$\mathbf{T} \equiv \begin{bmatrix} 1 & 1 & 2 \\ -2 & 3 & 5 \\ 4 & -2 & 5 \end{bmatrix}$$

Determine the vectors $\mathbf{T(e_1)}$ and $\mathbf{T(2e_1 - e_2 + e_3)}$.

1.73 Let $\mathbf{e_1}$ and $\mathbf{e_2}$ be two unit vectors along the coordinate axes x and y in two-dimensional space. Let $\mathbf{R}$ be a transformation that rotates every vector counterclockwise by an angle θ.

(a) Show that $\mathbf{R}$ is a linear transformation.

(b) Determine the 2×2 matrix representation of $\mathbf{R}$.

1.74 Solve Problem 1.21 again using equations (1.33) and (1.35).

1.75 Solve Problem 1.22 again using equations (1.33) and (1.35).

1.6 Cartesian Tensors

A linear transformation as defined in Section 1.5 is called a second-order tensor. In this section we will study general tensors including second-order tensors. The discussion will be limited to three-dimensional rectangular coordinates and Cartesian tensors. Maple has a special package called the *Tensor Package* that is invoked with the following command:

```
>  with(tensor);
```

[*Christoffel1*, *Christoffel2*, *Einstein*, *Jacobian*, *Killing_eqns*, *Levi_Civita*, *Lie_diff*, *Ricci*, *Ricciscalar*, *Riemann*, *RiemannF*, *Weyl*, *act*, *antisymmetrize*, *change_basis*, *commutator*, *compare*, *conj*, *connexF*, *contract*, *convertNP*, *cov_diff*, *create*, *d1metric*, *d2metric*, *directional_diff*, *displayGR*, *display_allGR*, *dual*, *entermetric*, *exterior_diff*, *exterior_prod*, *frame*, *geodesic_eqns*, *get_char*, *get_compts*, *get_rank*, *init*, *invars*, *invert*, *lin_com*, *lower*, *npcurve*, *npspin*, *partial_diff*, *permute_indices*, *petrov*, *prod*, *raise*, *symmetrize*, *tensorsGR*, *transform*]

Since second-order tensors can be represented by matrices, the *Linear Algebra Package* in Maple can be used also for second-order tensors. However, the *Tensor Package* should be used for general tensors.

An orthogonal tensor is a tensor whose inverse is equal to its transpose (see Example 1.11). Let $\mathbf{Q}$ be an orthogonal second-order tensor given by the following transformation:

$$\mathbf{e_i'} = Q_{mi}\mathbf{e_m} \tag{1.47}$$

where $\mathbf{e_i'}$ is a transformed coordinate system that can be obtained from $\mathbf{e_m}$ by a combination of a rotation and a reflection (see Figure 1.9).

Equation (1.47) can be represented by the following matrix equation:

$$\begin{bmatrix} \mathbf{e_1'} \\ \mathbf{e_2'} \\ \mathbf{e_3'} \end{bmatrix} = \begin{bmatrix} Q_{11} & Q_{12} & Q_{13} \\ Q_{12} & Q_{22} & Q_{23} \\ Q_{13} & Q_{23} & Q_{33} \end{bmatrix} \begin{bmatrix} \mathbf{e_1} \\ \mathbf{e_2} \\ \mathbf{e_3} \end{bmatrix} \tag{1.48}$$

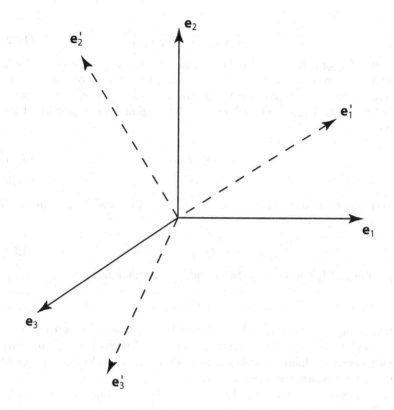

FIGURE 1.9
Coordinate systems $\mathbf{e_m}$ and $\mathbf{e'_i}$.

Since the matrix $\mathbf{Q}$ is orthogonal, we have $\mathbf{Q}^{-1} = \mathbf{Q}^T$. Therefore, we have the following relation:

$$\mathbf{Q}\mathbf{Q}^T = \mathbf{Q}^T\mathbf{Q} = \mathbf{I} \tag{1.49}$$

Equation (1.49) can be written using indicial notations as follows:

$$Q_{im}Q_{jm} = Q_{mi}Q_{mj} = \delta_{ij} \tag{1.50}$$

The components of $\mathbf{Q}$ are obtained using equation (1.42) as follows:

$$Q_{mi} = \mathbf{Q}(\mathbf{e_i}) \cdot \mathbf{e_m} \tag{1.51}$$

However, $\mathbf{Q}(\mathbf{e_i}) = \mathbf{e'_i}$ (see equation (1.47)). Therefore, equation (1.51) becomes:

$$Q_{mi} = \mathbf{e}_i' \cdot \mathbf{e_m} = \cos(\mathbf{e}_i', \mathbf{e_m}) \tag{1.52}$$

where $\cos(\mathbf{e}_i', \mathbf{e_m})$ is the cosine of the angle between the vectors $\mathbf{e}_i'$ and $\mathbf{e_m}$.

We will now consider how the components of a general vector $\mathbf{v}$ in the coordinate system $\mathbf{e_m}$ transform to the coordinate system $\mathbf{e}_i'$. The components of $\mathbf{v}$ can be written in each coordinate system using equation (1.38) as follows:

$$v_m = \mathbf{v} \cdot \mathbf{e_m} \tag{1.53a}$$

$$v_i' = \mathbf{v} \cdot \mathbf{e}_i' \tag{1.53b}$$

Substituting equation (1.47) into equation (1.53b) and comparing with equation (1.53a), we obtain:

$$v_i' = Q_{mi}v_m \tag{1.54}$$

Equation (1.54) can be written in matrix form as follows:

$$\mathbf{v}' = \mathbf{Q^T v} \tag{1.55}$$

When using equation (1.55), it should be noted that $\mathbf{v}$ is the vector with respect to the coordinate system $\mathbf{e_m}$ while $\mathbf{v}'$ is the same vector with respect to the coordinate system $\mathbf{e}_i'$. Equations (1.54) and (1.55) represent the general transformation law for any vector $\mathbf{v}$.

Next, we will consider how the components of a general second-order tensor $\mathbf{T}$ in the coordinate system $\mathbf{e_m}$ transform to the coordinate system $\mathbf{e}_i'$. The components of $\mathbf{T}$ can be written in each coordinate system using equation (1.42) as follows:

$$T_{mn} = T(\mathbf{e_n}) \cdot \mathbf{e_m} \tag{1.56a}$$

$$T_{ij}' = T(\mathbf{e}_j') \cdot \mathbf{e}_i' \tag{1.56b}$$

Substituting equation (1.47) into equation (1.56b) we obtain:

$$T_{ij}' = T(Q_{nj} \cdot \mathbf{e_n}) \cdot Q_{mi}\mathbf{e_m} = Q_{mi}Q_{nj}T(\mathbf{e_n}) \cdot \mathbf{e_m} \tag{1.57}$$

Comparing equations (1.56a) and (1.57), we obtain:

$$T_{ij}' = Q_{mi}Q_{nj}T_{mn} \tag{1.58}$$

Equation (1.58) can be written in matrix form as follows:

$$\mathbf{T}' = \mathbf{Q^T T Q} \tag{1.59}$$

When using equation (1.59), it should be noted that $\mathbf{T}'$ represents the tensor with respect to the coordinate system $\mathbf{e}_i'$ (or $\mathbf{e}_j'$), while $\mathbf{T}$ represents the same tensor with respect to the coordinate system $\mathbf{e_m}$ (or $\mathbf{e_n}$). Equations (1.58) and (1.59) represent the general transformation law for any second-order tensor $\mathbf{T}$.

In general, a scalar c is considered a zeroth-order tensor with the transformation law:

$$c' = c \tag{1.60a}$$

A vector $\mathbf{v}$ is considered a first-order tensor with the transformation law:

$$v_i' = Q_{mi}v_m \tag{1.60b}$$

A second-order tensor $\mathbf{T}$ (usually called just a tensor) has the transformation law:

$$T_{ij}' = Q_{mi}Q_{nj}T_{mn} \tag{1.60c}$$

Continuing along the same line, a third-order tensor $\mathbf{L}$ has the transformation law:

$$L_{ijk}' = Q_{mi}Q_{nj}Q_{pk}L_{mnp} \tag{1.60d}$$

A fourth-order tensor $\mathbf{M}$ will have the following transformation law:

$$M_{ijkl}' = Q_{mi}Q_{nj}Q_{pk}Q_{ql}M_{mnpq} \tag{1.60e}$$

Higher-order tensors can be defined using a transformation law along the lines of equations (1.60).

Example 1.11

Let $\mathbf{Q}$ be an orthogonal second-order tensor. Show that for any two general vectors $\mathbf{u}$ and $\mathbf{v}$, the following relation holds:

$$\mathbf{Q(u)} \cdot \mathbf{Q(v)} = \mathbf{u} \cdot \mathbf{v}$$

Solution

Since $\mathbf{Q}$ is orthogonal, we have:

$$\mathbf{Q^T Q = I}$$

$$\mathbf{I - Q^T Q = 0}$$

$$\mathbf{(I - Q^T Q)(u) = 0}$$

$$\mathbf{v} \cdot (\mathbf{I} - \mathbf{Q}^{\mathbf{T}}\mathbf{Q})(\mathbf{u}) = 0$$

$$\mathbf{v} \cdot \mathbf{I}(\mathbf{u}) = \mathbf{v} \cdot (\mathbf{Q}^{\mathbf{T}}\mathbf{Q})(\mathbf{u})$$

$$\mathbf{v} \cdot \mathbf{u} = \mathbf{v} \cdot \mathbf{Q}^{\mathbf{T}}(\mathbf{Q}(\mathbf{u}))$$

$$\mathbf{v} \cdot \mathbf{u} = \mathbf{Q}(\mathbf{v}) \cdot \mathbf{Q}(\mathbf{u})$$

Example 1.12

Let $\mathbf{Q}$ be an orthogonal second-order tensor. Show that $|\mathbf{Q}| = \pm 1$, where $|\mathbf{Q}|$ is the determinant of $\mathbf{Q}$.

Solution

Since $\mathbf{Q}$ is orthogonal, we have:

$$\mathbf{Q}\mathbf{Q}^{\mathbf{T}} = \mathbf{I}$$

$$|\mathbf{Q}\mathbf{Q}^{\mathbf{T}}| = |\mathbf{I}|$$

$$|\mathbf{Q}||\mathbf{Q}^{\mathbf{T}}| = 1$$

but

$$|\mathbf{Q}| = |\mathbf{Q}^{\mathbf{T}}|$$

$$|\mathbf{Q}||\mathbf{Q}| = 1$$

$$|\mathbf{Q}|^2 = 1$$

$$|\mathbf{Q}| = \pm 1$$

Example 1.13

Suppose that $\mathbf{v}$ is a vector and $\mathbf{T}$ is a second-order tensor related by the equation

$$v_i = A_{ijk}T_{jk} \tag{1.61}$$

Show that A_{ijk} are the components of a third-order tensor.

Solution

Since **v** is a vector, it transforms according to the relation (see equation (1.60b)):

$$v_i' = Q_{mi}v_m \qquad (1.62a)$$

Pre-multiply both sides of equation (1.62a) by Q_{ni} and noting that $Q_{ni}Q_{mi} = \delta_{nm}$, we obtain:

$$Q_{ni}v_i' = \delta_{nm}v_m = v_n \qquad (1.62b)$$

Switching indices in equation (1.62b), we obtain:

$$v_i = Q_{in}v_n' \qquad (1.63)$$

Since **T** is a second-order tensor, we have the following relation (see equation (1.60c)):

$$T_{ij}' = Q_{mi}Q_{ni}T_{mn} \qquad (1.64a)$$

Pre-multiplying equation (1.64a) by $Q_{pi}Q_{qj}$, we obtain:

$$
\begin{aligned}
Q_{pi}Q_{qj}T_{ij}' &= Q_{pi}Q_{qj}Q_{mi}Q_{nj}T_{mn} \\
&= \delta_{pm}\delta_{qn}T_{mn} \\
&= T_{pq} \qquad (1.64b)
\end{aligned}
$$

Switching indices (using jk for pq and mn for ij) in equation (1.64b), we obtain:

$$T_{jk} = Q_{jm}Q_{kn}T_{mn}' \qquad (1.65)$$

Next, substitute equations (1.63) and (1.65) into equation (1.61) to obtain:

$$Q_{in}v_n' = A_{ijk}Q_{jm}Q_{kn}T_{mn}' \qquad (1.66a)$$

Pre-multiply equation (1.66a) by Q_{ir}:

$$Q_{ir}Q_{in}v_n' = Q_{ir}Q_{jm}Q_{kn}A_{ijk}T_{mn}' \qquad (1.66b)$$

Substituting $Q_{ir}Q_{in} = \delta_{rn}$, equation (1.66b) becomes:

$$
\begin{aligned}
v_r' &= (Q_{ir}Q_{jm}Q_{kn}A_{ijk})T_{mn}' \\
&= A_{rmn}'T_{mn}' \qquad (1.67)
\end{aligned}
$$

where

$$A'_{rmn} = Q_{ir}Q_{jm}Q_{kn}A_{ijk} \tag{1.68}$$

It should now be clear from equation (1.68) that A_{ijk} are the components of a third order tensor (see equation (1.60d)).

Example 1.14

A second-order tensor $\mathbf{T}$ is *symmetric* if $T_{ij} = T_{ji}$, or equivalently, $\mathbf{T} = \mathbf{T^T}$. On the other hand, a tensor $\mathbf{T}$ is *antisymmetric* if $T_{ij} = -T_{ji}$, or equivalently, $\mathbf{T} = -\mathbf{T^T}$. Suppose that three second-order tensors $\mathbf{A}$, $\mathbf{B}$, and $\mathbf{C}$ are related by $\mathbf{C} = \mathbf{A} + \mathbf{B}$, where

$$\mathbf{A} = \frac{1}{2}(\mathbf{C} + \mathbf{C}^T) \tag{1.69a}$$

$$\mathbf{B} = \frac{1}{2}(\mathbf{C} - \mathbf{C}^T) \tag{1.69b}$$

Show that $\mathbf{A}$ is symmetric tensor (called the symmetric part of $\mathbf{C}$) and $\mathbf{B}$ is an antisymmetric tensor (called the antisymmetric part of $\mathbf{C}$).

Solution

Determine the transpose of both $\mathbf{A}$ and $\mathbf{B}$ as follows:

$$\mathbf{A^T} = \frac{1}{2}(\mathbf{C} + \mathbf{C^T})^{\mathbf{T}} = \frac{1}{2}(\mathbf{C^T} + (\mathbf{C^T})^{\mathbf{T}})$$
$$= \frac{1}{2}(\mathbf{C^T} + \mathbf{C}) = \frac{1}{2}(\mathbf{C} + \mathbf{C^T})$$

$$= \mathbf{A}$$

Therefore, $\mathbf{A}$ is a symmetric tensor.

$$\mathbf{B^T} = \frac{1}{2}(\mathbf{C} - \mathbf{C^T})^{\mathbf{T}} = \frac{1}{2}(\mathbf{C^T} - (\mathbf{C^T})^{\mathbf{T}})$$
$$= \frac{1}{2}(\mathbf{C^T} - \mathbf{C}) = -\frac{1}{2}(\mathbf{C} - \mathbf{C^T})$$
$$= -\mathbf{B}$$

Therefore, $\mathbf{B}$ is an antisymmetric tensor.

Problems

1.76 Let **T** be a second-order tensor that has the following matrix representation with respect to e_1, e_2, and e_3:

$$\mathbf{T} \equiv \begin{bmatrix} 5 & -2 & 0 \\ 0 & 1 & 4 \\ 3 & 2 & 3 \end{bmatrix}$$

Suppose that the primed coordinate axes e_1, e_2, and e_3 are obtained by rotating every vector about e_3 by an angle of $45°$ counterclockwise. Determine the matrix representation **T'** of **T** in the primed coordinate axes.

1.77 Let **A** and **B** be two second-order tensors related by the equation $B_{ij} = C_{ijkl}A_{kl}$. Show that C_{ijkl} are the components of a fourth-order tensor.

1.78 Let T_{ij} be the components of a second-order tensor **T**. Show that T_{ii} is a *scalar invariant*, i.e., show that $T_{ii} = T'_{mm}$ with respect to any orthogonal transformation.

1.79 Check the result given in Problem 1.78 by using the example given in Problem 1.76, i.e., determine both T_{ii} and T'_{mm} and show that they are equal.

1.80 Let T_{ij} be the components of a second-order tensor **T**. Show that $T_{ij}T_{ij}$ is a scalar invariant, i.e., $T_{ij}T_{ij} = T'_{mn}T'_{mn}$ for any orthogonal transformation.

1.81 Check the results given in Problem 1.80 by using the following matrix representations for the tensor **T** and the orthogonal transformation **Q**:

$$\mathbf{T} \equiv \begin{bmatrix} 1 & 0 & -2 \\ 5 & 2 & 4 \\ 3 & 1 & 3 \end{bmatrix}$$

$$\mathbf{Q} \equiv \begin{bmatrix} 1 & 0 & 0 \\ 1 & 0 & 1 \\ 0 & 1 & 0 \end{bmatrix}$$

where $e'_i = \mathbf{Q}(e'_i)$. Determine both $T_{ij}T_{ij}$ and $T'_{mn}T'_{mn}$ and show that they are equal.

1.82 Let **T** be a second-order tensor represented by the following matrix:

$$\mathbf{T} \equiv \begin{bmatrix} 3 & 1 & 10 \\ -2 & 5 & 0 \\ 3 & 5 & 7 \end{bmatrix}$$

Determine the matrix representations of both the symmetric part and the antisymmetric part of **T**. (*Hint*: Use Example 1.14)

1.83 Let E_{ijkl} be the components of a fourth-order tensor. Show that E_{ijmm} are the components of a second-order tensor.

1.84 Let A_{ij} and B_{ij} be the components of second-order tensors **A** and **B**, respectively. Show that the product $A_{im}B_{mj}$ are the components of a second-order tensor.

1.85 Let $T_{ijpq} = \alpha\delta_{ij}\delta_{pq} + \beta\delta_{ip}\delta_{jq}$ where α and β are scalars. Show that T_{ijpq} are the components of a fourth-order tensor.

1.86 Let **R** be a symmetric second-order tensor that is nonsingular, i.e., it has an inverse $\mathbf{R}^{-1}$. Show that $\mathbf{R}^{-1}$ is also symmetric.

1.87 The *dyadic product* **uv** of two vectors **u** and **v** is defined by $(\mathbf{uv})(\mathbf{w}) = (\mathbf{v} \cdot \mathbf{w})\mathbf{u}$ for any vector **w**.

   (a) Show that **uv** is a second order tensor.
   (b) Show that $(\mathbf{uv})_{ij} = u_i v_j$.
   (c) Determine the matrix representation of **uv**.

1.88 The transpose $\mathbf{T}^{\mathbf{T}}$ of a second-order tensor **T** is defined by the relation $\mathbf{u} \cdot \mathbf{T}(\mathbf{v}) = \mathbf{v} \cdot \mathbf{T}^{\mathbf{T}}(\mathbf{u})$ for any vectors **u** and **v**. Show that $(\mathbf{PQ})^T = \mathbf{Q}^{\mathbf{T}}\mathbf{P}^{\mathbf{T}}$ for any two second-order tensors **P** and **Q**.

1.89 Let **A** be an antisymmetric tensor. We can find a vector **u** such that $\mathbf{A}(\mathbf{v}) = \mathbf{u} \times \mathbf{v}$ for any arbitrary vector **v**. The vector **u** is called the *dual vector* or the *axial vector* of **A**. Show that $\mathbf{u} = -\frac{1}{2}\epsilon_{ijk}A_{jk}\mathbf{e_i}$.

1.90 Let **T** be a second-order tensor represented by the following matrix:

$$\mathbf{T} = \begin{bmatrix} 4 & 1 & 1 \\ -3 & 2 & 5 \\ 4 & 1 & 0 \end{bmatrix}$$

(a) Determine the antisymmetric part of $\mathbf{T}$, denoted by $\mathbf{T^A}$.

(b) Determine the axial vector of $\mathbf{T^A}$ (see Problem 1.89).

(c) Check that $\mathbf{T^A(v)} = \mathbf{u} \times \mathbf{v}$ for $\mathbf{v} = \mathbf{e_1} + \mathbf{e_2} + \mathbf{e_3}$.

1.91 Let $\mathbf{u}$ and $\mathbf{v}$ be two vectors. Show that $u_i v_i$ is a scalar invariant, i.e., show that $u_i v_i = u'_m v'_m$ for any orthogonal transformation.

1.92 Show that the determinant of a matrix is a scalar invariant, i.e., show that $|\mathbf{T}| = |\mathbf{T'}|$ where $\mathbf{T}$ is the matrix of tensor $\mathbf{T}$ with respect to $\mathbf{e_1, e_2}$, and $\mathbf{e_3}$, and $\mathbf{T'}$ is the matrix of $\mathbf{T}$ with respect to $\mathbf{e'_1, e'_2}$, and $\mathbf{e'_3}$.

1.93 Let $\mathbf{T}$ be a second-order tensor and let $\mathbf{A}$ and $\mathbf{B}$ be the symmetric and antisymmetric parts of $\mathbf{T}$, respectively. Show that $\mathbf{u \cdot T(u)} = \mathbf{u \cdot A(u)}$ for any vector $\mathbf{u}$, i.e., $\mathbf{u \cdot B(u)} = 0$.

1.94 Let $\mathbf{S}$ be a symmetric second-order tensor. Show that $\epsilon_{ijk} S_{jk} = 0$ where ϵ_{ijk} is the permutation symbol.

1.95 An isotropic tensor is a tensor which has the same set of components in all coordinate systems.

(a) Show that all scalars are isotropic tensors.

(b) Any second-order isotropic tensor can be written as $\alpha \mathbf{I}$ where α is a scalar and $\mathbf{I}$ is the identity matrix.

1.96 Show that there are no nontrivial first-order isotropic tensors (use the definition in Problem 1.95).

1.97 Show that any fourth-order isotropic tensor $\mathbf{T}$ can be written in component form as follows:

$$T_{ijmn} = \alpha(\delta_{ij}\delta_{mn}) + \beta(\delta_{im}\delta_{jn} + \delta_{in}\delta_{jm}) + \gamma(\delta_{im}\delta_{jn} - \delta_{in}\delta_{jm})$$

where α, β, and γ are scalars (use the definition in Problem 1.95).

1.98 Let M_{ij} and N_{ij} be components of second-order tensors $\mathbf{M}$ and $\mathbf{N}$, respectively. Suppose M_{ij} and N_{ij} are related by the relation $M_{ij} = P_{ijmn}N_{mn}$. Show that if M_{ij} is symmetric, then $P_{ijmn} = P_{jimn}$.

1.99 Let $\mathbf{M}$, $\mathbf{N}$, and $\mathbf{P}$ be related by the same relation given in Problem 1.98, i.e., $M_{ij} = P_{ijmn}N_{mn}$. Show that if $\mathbf{N}$ is symmetric and $\mathbf{P}$ is isotropic, then $M_{ij} = 2\alpha N_{ij} + \beta N_{kk}\delta_{ij}$ where α and β are scalars.

1.7   Eigenvalues and Eigenvectors of Tensors

Eigenvalues and eigenvectors are defined for second-order tensors in a similar way to their definition for matrices as described in section 1.3. Therefore, for any second-order tensor $\mathbf{T}$, the scalar λ is called an eigenvalue of the tensor if $\mathbf{T(V)} = \lambda\mathbf{v}$ for any arbitrary vector $\mathbf{v}$. The vector $\mathbf{v}$ is called an eigenvector of $\mathbf{T}$. Usually eigenvectors are normalized as unit vectors.

Let $\mathbf{n}$ be a unit eigenvector of the tensor $\mathbf{T}$, then $\mathbf{T(n)} = \lambda\mathbf{n} = \lambda\mathbf{I(n)}$. Therefore, we have the following equation:

$$(\mathbf{T} - \lambda\mathbf{I})\mathbf{n} = \mathbf{0} \qquad (1.70)$$

Equation (1.70) can be solved only if the determinant of the tensor $\mathbf{T} - \lambda\mathbf{I}$ vanishes:

$$|\mathbf{T} - \lambda\mathbf{I}| = 0 \qquad (1.71)$$

For a second-order tensor $\mathbf{T}$, equation (1.71) is a cubic polynomial equation in λ – called the characteristic polynomial of $\mathbf{T}$. The three roots of the characteristic polynomial are the eigenvalues of the tensor $\mathbf{T}$.

The eigenvalues of a tensor are also called the *principal values* of the tensor, while the eigenvectors are called the *principal directions* of the tensor. The characteristic polynomial of a tensor $\mathbf{T}$ as given in equation (1.71) can be written explicitly as follows:

$$\lambda^3 - I_1\lambda^2 + I_2\lambda - I_3 = 0 \qquad (1.72)$$

where I_1, I_2, and I_3 are called the *scalar invariants* of $\mathbf{T}$, given by:

$$
\begin{aligned}
I_1 &= T_{ii} \\
I_2 &= \frac{1}{2}\left(T_{ii}T_{jj} - T_{ij}T_{ji}\right) \\
I_3 &= |\mathbf{T}|
\end{aligned}
\qquad (1.73)
$$

It should be noted that the scalar invariants of $\mathbf{T}$ remain unchanged regardless of the base vectors or coordinate system used.

Example 1.15

Let $\mathbf{T}$ be a second-order real symmetric tensor. Show that if the eigenvalues of $\mathbf{T}$ are all distinct, then the principal directions of $\mathbf{T}$ are mutually perpendicular.

Solution

Let λ_1 and λ_2 be two distinct eigenvalues (i.e. $\lambda_1 \neq \lambda_2$) of $\mathbf{T}$ with the corresponding eigenvectors $\mathbf{n_1}$ and $\mathbf{n_2}$. Then we have the following two equations:

$$\mathbf{T(n_1)} = \lambda_1 \mathbf{n_1}$$
$$\mathbf{T(n_2)} = \lambda_2 \mathbf{n_2} \tag{1.74}$$

Multiply (dot product) the first equation of (1.74) with $\mathbf{n_2}$ and multiply (dot product) the second equation of (1.74) with $\mathbf{n_1}$ to obtain:

$$\mathbf{n_2} \cdot \mathbf{T(n_1)} = \mathbf{n_2} \cdot \lambda_1 \mathbf{n_1}$$
$$\mathbf{n_1} \cdot \mathbf{T(n_2)} = \mathbf{n_1} \cdot \lambda_2 \mathbf{n_2} \tag{1.75}$$

However, $\mathbf{n_1} \cdot \mathbf{T(n_2)} = \mathbf{n_2} \cdot \mathbf{T}^T(\mathbf{n_1}) = \mathbf{n_2} \cdot \mathbf{T(n_1)}$ since $\mathbf{T}$ is symmetric. Therefore, the two equations (1.75) are identical so that we have:

$$\mathbf{n_2} \cdot \lambda_1 \mathbf{n_1} = \mathbf{n_1} \cdot \lambda_2 \mathbf{n_2} \tag{1.76}$$

Equation (1.76) can be re-written as follows:

$$(\lambda_1 - \lambda_2)(\mathbf{n_1} \cdot \mathbf{n_2}) = 0 \tag{1.77}$$

Since $\lambda_1 \neq \lambda_2$, we conclude that $\mathbf{n_1} \cdot \mathbf{n_2} = 0$ and thus the two eigenvectors $\mathbf{n_1}$ and $\mathbf{n_2}$ are mutually perpendicular.

Example 1.16

Let $\mathbf{T}$ be a second-order real symmetric tensor and $\mathbf{n_1}$, $\mathbf{n_2}$, and $\mathbf{n_3}$ be three unit vectors along the principal directions of $\mathbf{T}$, which are mutually perpendicular. Let $\lambda_1, \lambda_2, \lambda_3$ be the corresponding eigenvalues.

(a) Show that the tensor $\mathbf{T}$ can be represented along the principal directions with a diagonal matrix of the form:

$$T \equiv \begin{bmatrix} \lambda_1 & 0 & 0 \\ 0 & \lambda_2 & 0 \\ 0 & 0 & \lambda3 \end{bmatrix}$$

(b) Show that, in this case, the scalar invariants of **T** are given by:

$$I_1 = \lambda_1 + \lambda_2 + \lambda_3$$
$$I_2 = \lambda, \lambda_2 + \lambda_2\lambda_3 + \lambda_3\lambda_1$$
$$I_3 = \lambda_1\lambda_2\lambda_3$$

Solution

(a) In order to determine the matrix representation of **T**, we use equation (1.42) as follows:

$$T_{11} = \mathbf{T(n_1)} \cdot \mathbf{n_1} = \lambda_1\mathbf{n_1} \cdot \mathbf{n_1} = \lambda_1$$
$$T_{12} = \mathbf{T(n_2)} \cdot \mathbf{n_1} = \lambda_2\mathbf{n_2} \cdot \mathbf{n_1} = 0$$
$$T_{13} = \mathbf{T(n_3)} \cdot \mathbf{n_1} = \lambda_3\mathbf{n_3} \cdot \mathbf{n_1} = 0$$
$$T_{21} = \mathbf{T(n_1)} \cdot \mathbf{n_2} = \lambda_1\mathbf{n_1} \cdot \mathbf{n_2} = 0$$
$$T_{22} = \mathbf{T(n_2)} \cdot \mathbf{n_2} = \lambda_2\mathbf{n_2} \cdot \mathbf{n_2} = \lambda_2$$
$$T_{23} = \mathbf{T(n_3)} \cdot \mathbf{n_2} = \lambda_3\mathbf{n_3} \cdot \mathbf{n_2} = 0$$
$$T_{31} = \mathbf{T(n_1)} \cdot \mathbf{n_3} = \lambda_1\mathbf{n_1} \cdot \mathbf{n_3} = 0$$
$$T_{32} = \mathbf{T(n_2)} \cdot \mathbf{n_3} = \lambda_2\mathbf{n_2} \cdot \mathbf{n_3} = 0$$
$$T_{33} = \mathbf{T(n_3)} \cdot \mathbf{n_3} = \lambda_3\mathbf{n_3} \cdot \mathbf{n_3} = \lambda_3$$

Therefore, we have;

$$T \equiv \begin{bmatrix} \lambda_1 & 0 & 0 \\ 0 & \lambda_2 & 0 \\ 0 & 0 & \lambda_3 \end{bmatrix}$$

(b) Using equations (1.73) for the scalar invariants and applying them to the diagonal matrix above, we obtain:

$$I_1 = T_{ii} = \lambda_1 + \lambda_2 + \lambda_3$$
$$I_2 = \tfrac{1}{2}\left(T_{ii}T_{jj} - T_{ij}T_{ji}\right)$$

First calculate $T_{ij}T_{ji}$:

$$\begin{aligned}
T_{ij}T_{ji} &= T_{1j}T_{j1} + T_{2j}T_{j2} + T_{3j}T_{j3} \\
&= T_{11}T_{11} + T_{12}T_{21} + T_{13}T_{31} \\
&\quad + T_{21}T_{12} + T_{22}T_{22} + T_{23}T_{32} \\
&\quad + T_{31}T_{13} + T_{32}T_{23} + T_{33}T_{33} \\
&= \lambda_1^2 + \lambda_2^2 + \lambda_3^2
\end{aligned}$$

$$I_2 = \frac{1}{2} \left[(\lambda_1 + \lambda_2 + \lambda_3)(\lambda_1 + \lambda_2 + \lambda_3) - (\lambda_1^2 + \lambda_2^2 + \lambda_3^2) \right]$$

$$= \frac{1}{2} \left(\lambda_1^2 + 2\lambda_1\lambda_2 + 2\lambda_1\lambda_3 + \lambda_2^2 + 2\lambda_2\lambda_3 + \lambda_3^2 - \lambda_1^2 - \lambda_2^2 - \lambda_3^2 \right)$$

$$= \lambda_1\lambda_2 + \lambda_2\lambda_3 + \lambda_3\lambda_1$$

$$I_3 = |\mathbf{T}| = \lambda_1 \begin{vmatrix} \lambda_2 & 0 \\ 0 & \lambda_3 \end{vmatrix} - 0 + 0$$

$$= \lambda_1\lambda_2\lambda_3 \tag{1.78}$$

Example 1.17

Let $\mathbf{T}$ be a second-order real symmetric tensor represented by the following matrix with respect to some particular coordinate system:

$$\mathbf{T} \equiv \begin{bmatrix} 3 & -1 & 0 \\ -1 & 2 & 5 \\ 0 & 5 & 2 \end{bmatrix}$$

Determine the eigenvalues and the corresponding eigenvectors of $\mathbf{T}$.

Solution

Applying equation (1.71), we obtain:

$$|\mathbf{T} - \lambda\mathbf{I}| = \begin{vmatrix} 3 - \lambda & -1 & 0 \\ -1 & 2 - \lambda & 5 \\ 0 & 5 & 2 - \lambda \end{vmatrix} = 0$$

$$(3 - \lambda) \begin{vmatrix} 2 - \lambda & 5 \\ 5 & 2 - \lambda \end{vmatrix} + 1 \begin{vmatrix} -1 & 5 \\ 0 & 2 - \lambda \end{vmatrix} = 0$$

$$(3 - \lambda)[(2 - \lambda)^2 - 25] + \lambda - 2 = 0$$

Simplifying the above equation, we obtain:

$$\lambda^3 - 7\lambda^2 - 10\lambda + 66 = 0$$

The three roots of the above cubic equation are the eigenvalues of $\mathbf{T}$ given by:

$$\lambda_1 = -3.1$$
$$\lambda_2 = -3.0$$
$$\lambda_3 = -7.1$$

The eigenvectors are then obtained by substituting each λ obtained into equation (1.70) and solving for $\mathbf{n}$ in each case. Therefore, the three eigenvectors are obtained as follows:

$$n_1 \equiv [0.1164 \, 0.7080 - 0.6965]^T$$

$$n_2 \equiv [-0.9784 - 0.0390 - 0.2032]^T$$

$$n_3 \equiv [0.1710 - 0.7051 - 0.6882]^T$$

Problems

1.100 Determine the eigenvalues and the corresponding eigenvectors for the second-order tensor $\mathbf{T}$ which is represented by the following matrix with respect to some particular coordinate system:

$$\mathbf{T} \equiv \begin{bmatrix} 1 & 0 & 1 \\ 0 & 2 & 0 \\ 1 & 0 & 3 \end{bmatrix}$$

1.101 Let $\mathbf{T}$ be a second-order real and symmetric tensor. Show that the eigenvalues of $\mathbf{T}$ include the maximum and minimum values that the diagonal elements of any matrix representation of $\mathbf{T}$ can have.

1.102 Consider the tensor $\mathbf{T}$ given in the Example 1.17. Determine the scalar invariants of $\mathbf{T}$ and find the eigenvalues using equation (1.72).

1.103 Let $\mathbf{T}$ be a second-order tensor represented by the following matrix with respect to some particular coordinate system:

$$\mathbf{T} \equiv \begin{bmatrix} 1 & 0 & 0 \\ 0 & 2 & 3 \\ 0 & 3 & -2 \end{bmatrix}$$

Can the tensor $\mathbf{T}$ be represented by the following matrix with respect to some other coordinate system (*Hint*: Use eigenvalues):

$$\begin{bmatrix} 1 & -1 & 0 \\ -1 & 0 & 5 \\ 0 & 5 & 2 \end{bmatrix}$$

1.104 Let $\mathbf{T}$ be a second-order tensor represented by the following matrix with respect to some particular coordinate system:

$$\mathbf{T} \equiv \begin{bmatrix} 1 & 2 & 2 \\ 2 & 3 & 5 \\ 2 & 5 & -1 \end{bmatrix}$$

Determine the diagonal matrix representation of $\mathbf{T}$ given in Example 1.16.

1.105 Show that the three scalar invariants I_1, I_2, and I_3 of equations (1.73) are scalars.

1.106 Determine the eigenvalues and eigenvectors of an antisymmetric tensor $\mathbf{T}$ represented by the following matrix with respect to some particular coordinate system:

$$\mathbf{T} \equiv \begin{bmatrix} 0 & 0 & 1 \\ 0 & 0 & 0 \\ -1 & 0 & 0 \end{bmatrix}$$

1.107 Consider the antisymmetric tensor $\mathbf{T}$ of Problem 1.106 above. Show that the eigenvectors corresponding to nonzero eigenvalues satisfy $\mathbf{v} \cdot \mathbf{v} = 0$. Show also that this property holds for all antisymmetric tensors.

1.108 Solve Example 1.17 again using Maple.

1.109 Solve Problem 1.100 again using Maple.

1.110 Solve Problem 1.103 again using Maple.

1.111 Solve Problem 1.104 again using Maple.

1.112 Solve Problem 1.106 again using Maple.

1.8 Tensor Calculus

In this section we study tensor functions of second-order tensors including tensor differentiation and integration. Let $\mathbf{T}(t)$ be a tensor-valued function of a scalar t. We will denote $\mathbf{T}(t)$ here by just $\mathbf{T}$. The tensor derivative $d\mathbf{T}/dt$ is a tensor defined by:

$$\frac{d\mathbf{T}}{dt} = \lim_{\Delta t \to 0} \frac{\mathbf{T}(t + \Delta t) - \mathbf{T}(t)}{\Delta t} \tag{1.79}$$

The following identities related to the tensor derivative can be easily proved:

$$\frac{d}{dt}(\mathbf{T} + \mathbf{S}) = \frac{d\mathbf{T}}{dt} + \frac{d\mathbf{S}}{dt} \tag{1.80a}$$

$$\frac{d}{dt}(\alpha(t)\mathbf{T}) = \frac{d\alpha}{dt}\mathbf{T} + \alpha\frac{d\mathbf{T}}{dt} \tag{1.80b}$$

$$\frac{d}{dt}(\mathbf{TS}) = \frac{d\mathbf{T}}{dt}\mathbf{S} + \mathbf{T}\frac{d\mathbf{S}}{dt} \tag{1.80c}$$

$$\frac{d}{dt}(\mathbf{T}(\mathbf{a})) = \frac{d\mathbf{T}}{dt}(\mathbf{a}) + \mathbf{T}\left(\frac{d\mathbf{a}}{dt}\right) \tag{1.80d}$$

$$\frac{d}{dt}(\mathbf{T}^T) = \left(\frac{d\mathbf{T}}{dt}\right)^T \tag{1.80e}$$

In the above equations, $\mathbf{T}$ and $\mathbf{S}$ are second-order tensors, $\mathbf{a}$ is a vector, and $\alpha(t)$ is a scalar function.

Let $\Phi(\mathbf{r})$ be a scalar-valued function of the position vector $\mathbf{r}$, i.e., $\Phi(\mathbf{r})$ is a scalar field. Define the *gradient* of Φ, defined as a vector, denoted by $\nabla\Phi$ as follows:

$$\nabla\Phi \cdot d\mathbf{r} = d\Phi = \Phi(\mathbf{r} + d\mathbf{r}) - \Phi(\mathbf{r}) \tag{1.81}$$

Dividing the above equation by dr (the magnitude or length of $d\mathbf{r}$), we obtain:

$$\left(\frac{d\Phi}{dr}\right)_{in\ \mathbf{e}-direction} = \nabla\Phi \cdot \mathbf{e} \tag{1.82}$$

where $\mathbf{e} = d\mathbf{r}/dr$ is a unit vector along $d\mathbf{r}$. Equation (1.81) means that the rate of change of Φ along $d\mathbf{r}$ (the directional derivative) is equal to the component of $\nabla\Phi$ in the direction of $\mathbf{e}$.

Applying equation (1.81) along each of the $\mathbf{e}_1$, $\mathbf{e}_2$, and $\mathbf{e}_3$ directions, we obtain:

$$\left(\frac{d\Phi}{dr}\right)_{in\ \mathbf{e}_1-direction} = \frac{\partial\Phi}{\partial x_1} = \nabla\Phi \cdot \mathbf{e}_1 = (\nabla\Phi)_1 \tag{1.83a}$$

$$\left(\frac{d\Phi}{dr}\right)_{in\ \mathbf{e}_2-direction} = \frac{\partial\Phi}{\partial x_2} = \nabla\Phi \cdot \mathbf{e}_2 = (\nabla\Phi)_2 \tag{1.83b}$$

$$\left(\frac{d\Phi}{dr}\right)_{in\ \mathbf{e_3}-direction} = \frac{\partial\Phi}{\partial x_3} = \nabla\Phi \cdot \mathbf{e_3} = (\nabla\Phi)_3 \qquad (1.83c)$$

Thus the vector $\nabla\Phi$ can be written in terms of its components as follows:

$$\nabla\Phi = \frac{\partial\Phi}{\partial \mathbf{x_i}}\mathbf{e_i} = \frac{\partial\Phi}{\partial \mathbf{x_1}}\mathbf{e_1} + \frac{\partial\Phi}{\partial \mathbf{x_2}}\mathbf{e_2} + \frac{\partial\Phi}{\partial \mathbf{x_3}}\mathbf{e_3} \qquad (1.84)$$

Geometrically, the gradient vector $bf\nabla\Phi$ is normal to the surface of constant Φ because in this case $d\Phi = 0$ for any dr tangent to the surface. Thus $\nabla\Phi \cdot d\mathbf{r} = 0$ and $\nabla\Phi$ is normal to the tangent to the surface.

Let $\mathbf{v}(\mathbf{r})$ be a vector-valued function of the position vector $\mathbf{r}$, i.e. $\mathbf{v}(\mathbf{r})$ is a vector field. We define the *gradient* of $\mathbf{v}$, a second-order tensor, denoted by $\nabla\mathbf{v}$ as follows:

$$(\nabla\mathbf{v})(d\mathbf{r}) = d\mathbf{v} = \mathbf{v}(\mathbf{r}+d\mathbf{r}) - \mathbf{v}(\mathbf{r}) \qquad (1.85)$$

Dividing the above equation by dr (the length of $d\mathbf{r}$), we obtain:

$$\left(\frac{d\mathbf{v}}{dr}\right)_{in\ \mathbf{e}-direction} = (\nabla\mathbf{v})(\mathbf{e}) \qquad (1.86)$$

where $\mathbf{e} = d\mathbf{r}/dr$ is a unit vector along $d\mathbf{r}$. Next, we will obtain the components of the gradient tensor $\nabla\mathbf{v}$ as follows:

$$(\nabla\mathbf{v})_{11} = \mathbf{e_1} \cdot (\nabla\mathbf{v})(\mathbf{e_1}) = \mathbf{e_1} \cdot \frac{\partial\mathbf{v}}{\partial x_1} = \frac{\partial}{\partial x_1}(\mathbf{e_1} \cdot \mathbf{v}) = \frac{\partial}{\partial x}v_1 = \frac{\partial v_1}{\partial x_1} \qquad (1.87)$$

In general,

$$(\nabla\mathbf{v})_{ij} = \mathbf{e_i} \cdot (\nabla\mathbf{v})(\mathbf{e_j}) = \mathbf{e_i} \cdot \frac{\partial\mathbf{v}}{\partial x_j} = \frac{\partial}{\partial x_j}(\mathbf{e_i} \cdot \mathbf{v}) = \frac{\partial v_i}{\partial x_j} \qquad (1.88)$$

Thus, the second-order gradient tensor $\nabla\mathbf{v}$ may be represented by the following matrix:

$$\nabla\mathbf{v} \equiv \begin{bmatrix} \frac{\partial v_1}{\partial x_1} & \frac{\partial v_1}{\partial x_2} & \frac{\partial v_1}{\partial x_3} \\ \frac{\partial v_2}{\partial x_1} & \frac{\partial v_2}{\partial x_2} & \frac{\partial v_2}{\partial x_3} \\ \frac{\partial v_3}{\partial x_1} & \frac{\partial v_3}{\partial x_2} & \frac{\partial v_3}{\partial x_3} \end{bmatrix} \qquad (1.89)$$

Let $\mathbf{v}(\mathbf{r})$ be a vector field as defined above. The *divergence* of $\mathbf{v}(\mathbf{r})$, denoted by div $\mathbf{v}$, is defined to be a scalar field as follows:

$$\text{div}\mathbf{v} = tr(\nabla\mathbf{v}) \qquad (1.90)$$

where tr is the trace of a second-order tensor. Substituting equation (1.88) into equation (1.89), we obtain:

$$\text{div}\mathbf{v} = \frac{\partial v_1}{\partial x_1} + \frac{\partial v_2}{\partial x_2} + \frac{\partial v_3}{\partial x_3} = \frac{\partial v_i}{\partial x_i} \tag{1.91}$$

Let $\mathbf{T}(\mathbf{r})$ be a second-order tensor-valued function of the position vector $\mathbf{r}$, i.e., $\mathbf{T}(\mathbf{r})$ is a tensor field. The *divergence* of $\mathbf{T}$, denoted by **div T**, is a vector field defined by:

$$(\mathbf{div}\mathbf{T}) \cdot \mathbf{a} = \text{div}(\mathbf{T}^T(\mathbf{a})) - tr(\mathbf{T}^T(\nabla \mathbf{a})) \tag{1.92}$$

Let $\mathbf{b} = \mathbf{div}\ \mathbf{T}$. To determine the components b_i of **div T**, we obtain:

$$\begin{aligned}
b_i = \mathbf{b} \cdot \mathbf{e_i} &= div\left(\mathbf{T}^T(\mathbf{e_i})\right) - tr\left(\mathbf{T}^T(\nabla \mathbf{e_i})\right) \\
&= div\left(T_{im}e_m\right) - tr\left(\mathbf{T}^T(0)\right) \\
&= div\left(T_{im}e_m\right) \\
&= \frac{\partial T_{im}}{\partial x_m}
\end{aligned} \tag{1.93}$$

Note in equation (1.92) that $\nabla \mathbf{e_i} = 0$ (see equation (1.87). Therefore, **div T** is a vector defined by:

$$\mathbf{div}\mathbf{T} = \frac{\partial T_{im}}{\partial x_m}\mathbf{e_i} \tag{1.94}$$

Let $\mathbf{v}(\mathbf{r})$ be a vector field as defined above. The *curl* of $\mathbf{v}$, denoted by **curlv**, is a vector field defined by:

$$\mathbf{curlv} = 2t^A \tag{1.95}$$

where t^A is the dual vector of the antisymmetric part of $\nabla \mathbf{v}$, i.e., $((\nabla \mathbf{v})^A$. See Problem 1.89 for the definition of the dual vector for an antisymmetric tensor.

Using equation (1.88), $(\nabla \mathbf{v})^A$ is obtained as follows:

$$(\nabla \mathbf{v})^A \equiv \begin{bmatrix}
0 & \frac{1}{2}\left(\frac{\partial v_1}{\partial x_2} - \frac{\partial v_2}{\partial x_1}\right) & \frac{1}{2}\left(\frac{\partial v_1}{\partial x_3} - \frac{\partial v_3}{\partial x_1}\right) \\
-\frac{1}{2}\left(\frac{\partial v_1}{\partial x_2} - \frac{\partial v_2}{\partial x_1}\right) & 0 & \frac{1}{2}\left(\frac{\partial v_2}{\partial x_3} - \frac{\partial v_3}{\partial x_2}\right) \\
-\frac{1}{2}\left(\frac{\partial v_1}{\partial x_3} - \frac{\partial v_3}{\partial x_1}\right) & -\frac{1}{2}\left(\frac{\partial v_2}{\partial x_3} - \frac{\partial v_3}{\partial x_2}\right) & 0
\end{bmatrix} \tag{1.96}$$

The curl of $\mathbf{v}$ is then obtained as follows:

$$\mathbf{curlv} = \left(\frac{\partial v_3}{\partial x_2} - \frac{\partial v_2}{\partial x_3}\right)\mathbf{e_1} + \left(\frac{\partial v_1}{\partial x_3} - \frac{\partial v_3}{\partial x_1}\right)\mathbf{e_2} + \left(\frac{\partial v_2}{\partial x_1} - \frac{\partial v_1}{\partial x_2}\right)\mathbf{e_3} \tag{1.97}$$

Example 1.18

Prove equation (1.79a)

Solution

$$
\begin{aligned}
\frac{d}{dt}(\mathbf{T}+\mathbf{S}) &= \lim_{\Delta t \to 0} \frac{(\mathbf{T}+\mathbf{S})(t+\Delta t) - (\mathbf{T}+\mathbf{S})(t)}{\Delta t} \\
&= \lim_{\Delta t \to 0} \frac{\mathbf{T}(t+\Delta t) + \mathbf{S}(t+\Delta t) - \mathbf{T}(t) - \mathbf{S}(t)}{\Delta t} \\
&= \lim_{\Delta t \to 0} \frac{\mathbf{T}(t+\Delta t) - \mathbf{T}(t)}{\Delta t} + \lim_{\Delta t \to 0} \frac{\mathbf{S}(t+\Delta t) - \mathbf{S}(t)}{\Delta t} \\
&= \frac{d\mathbf{T}}{dt} + \frac{d\mathbf{S}}{dt}
\end{aligned}
$$

Example 1.19

Given the scalar field $\Phi = x^2 + 2yz - 3z^2$, determine the vector $\mathbf{n} = -\alpha \nabla \Phi$ where α is a scalar.

Solution

$$
\begin{aligned}
\nabla \Phi &= \frac{\partial \Phi}{\partial x}\mathbf{e_1} + \frac{\partial \Phi}{\partial y}\mathbf{e_2} + \frac{\partial \Phi}{\partial z}\mathbf{e_3} \\
&= 2x\mathbf{e_1} + 2z\mathbf{e_2} + (2y - 6z)\mathbf{e_3} \\
\mathbf{n} &= -2\alpha[x\mathbf{e_1} + z\mathbf{e_2} + (y - 3z)\mathbf{e_3}]
\end{aligned}
$$

Example 1.20

Consider the vector field $\mathbf{v} = x^2\mathbf{e_1} + y^2\mathbf{e_2} + z^2\mathbf{e_3}$. Determine the matrix representation of the tensor $\nabla \mathbf{v}$ at the point $(1,1,1)$.

Solution

$$
\nabla \mathbf{v} \equiv
\begin{bmatrix}
\frac{\partial v_1}{\partial x} & \frac{\partial v_1}{\partial y} & \frac{\partial v_1}{\partial z} \\
\frac{\partial v_2}{\partial x} & \frac{\partial v_2}{\partial y} & \frac{\partial v_2}{\partial z} \\
\frac{\partial v_3}{\partial x} & \frac{\partial v_3}{\partial y} & \frac{\partial v_3}{\partial z}
\end{bmatrix}
$$

where

$$
\begin{aligned}
v_1 &= \mathbf{e_1} \cdot \mathbf{v} = x^2 \\
v_2 &= \mathbf{e_2} \cdot \mathbf{v} = y^2 \\
v_3 &= \mathbf{e_3} \cdot \mathbf{v} = z^2
\end{aligned}
$$

$$\nabla \mathbf{v} = \begin{bmatrix} 2x & 0 & 0 \\ 0 & 2y & 0 \\ 0 & 0 & 2z \end{bmatrix}$$

At the point (1,1,1),

$$\nabla \mathbf{v} = \begin{bmatrix} 2 & 0 & 0 \\ 0 & 2 & 0 \\ 0 & 0 & 2 \end{bmatrix} = 2\mathbf{I}$$

Example 1.21

Let Φ be a scalar field and $\mathbf{v}$ be a vector field. Prove the following relation:

$$div(\Phi\mathbf{v}) = \nabla\Phi \cdot \mathbf{v} + \Phi(divv)$$

Solution

$$
\begin{aligned}
ldiv(\Phi\mathbf{v}) &= tr(\nabla(\Phi\mathbf{v})) \\
&= \frac{\partial(\Phi v_1)}{\partial x_1} + \frac{\partial(\Phi v_2)}{\partial x_2} + \frac{\partial(\Phi v_3)}{\partial x_3} \\
&= \Phi\frac{\partial v_1}{\partial x_1} + \frac{\partial\Phi}{\partial x_1}v_1 + \Phi\frac{\partial v_2}{\partial x_2} + \frac{\partial\Phi}{\partial x_2}v_2 + \Phi\frac{\partial v_3}{\partial x_3} + \frac{\partial\Phi}{\partial x_3}v_3 \\
&= \left(\frac{\partial\Phi}{\partial x_1}v_1 + \frac{\partial\Phi}{\partial x_2}v_2 + \frac{\partial\Phi}{\partial x_3}v_3\right) + \left(\Phi\frac{\partial v_1}{\partial x_1} + \Phi\frac{\partial v_2}{\partial x_2} + \Phi\frac{\partial v_3}{\partial x_3}\right) \\
&= \nabla\Phi \cdot \mathbf{v} + \Phi(divv)
\end{aligned}
$$

Example 1.22

Consider the vector field $\mathbf{v}$ given in Example 1.20. Determine div $\mathbf{v}$ and **curl v** at the point (1,1,1).

Solution

$$div\mathbf{v} = \frac{\partial v_1}{\partial x} + \frac{\partial v_2}{\partial y} + \frac{\partial v_3}{\partial z} = 2 + 2 + 2 = 6$$

$$
\begin{aligned}
\mathbf{curl\ v} &= \left(\frac{\partial v_3}{\partial y} - \frac{\partial v_2}{\partial z}\right)\mathbf{e_1} + \left(\frac{\partial v_1}{\partial z} - \frac{\partial v_3}{\partial x}\right)\mathbf{e_2} + \left(\frac{\partial v_2}{\partial x} - \frac{\partial v_1}{\partial y}\right)\mathbf{e_3} \\
&= (0)\,e_1 + (0)\,e_2 + (0)\,e_3 \\
&= 0
\end{aligned}
$$

Problems

1.113 Prove equations (1.79b), (1.79c), (1.79d), and (1.79e).

1.114 Given the scalar field $\Psi(x, y, z) = 2xyz$, determine the gradient vector $\nabla\Psi$.

1.115 Given the vector field $\mathbf{v} = xy\mathbf{e_1} + xz\mathbf{e_2} + yz\mathbf{e_3}$, determine $\nabla\mathbf{v}$, div $\mathbf{v}$, and **curl v**.

1.116 For a general scalar field Φ, show that $\mathbf{curl}(\nabla\Phi) = \mathbf{0}$.

1.117 For a general vector field $\mathbf{v}$, show that $\mathrm{div}(\mathbf{curl\ v}) = 0$.

1.118 Let Φ and Ψ be two general scalar fields. Show that $\nabla(\Phi + \Psi) = \nabla\Phi + \nabla\Psi$.

1.119 Let $\mathbf{u}$ and $\mathbf{v}$ be two general vector fields. Show that $\mathrm{div}(\mathbf{u} + \mathbf{v}) = \mathrm{div}\ \mathbf{u} + \mathrm{div}\ \mathbf{v}$.

1.120 Prove equation (1.96).

1.121 For any general vector field $\mathbf{v}$, show that $\mathbf{curl}(\mathbf{curl v}) = \nabla(\mathrm{div}\ \mathbf{v}) - \nabla(\nabla\mathbf{v})$.

1.122 Show that $(\mathbf{curl v})_i = \epsilon_{ijk}\frac{\partial v_k}{\partial x_j}$ where $\mathbf{v}$ is a general vector field.

1.123 Define the Laplacian $\nabla^2\mathbf{v}$ of a vector field $\mathbf{v}$ as $\nabla^2 = \nabla(\nabla\mathbf{v})$. Show that $\left(\nabla^2\mathbf{v}\right)_j = \frac{\partial^2 v_j}{\partial x_i \partial x_i}$.

1.124 Let Φ and Ψ be two general scalar fields. Show that $\nabla(\Phi\Psi) = (\nabla\Phi)\Psi + \Phi(\nabla\Psi)$.

1.125 Let $\mathbf{v}$ be a vector field given by $\mathbf{v} = x^2 y\mathbf{e_1} + xy^2\mathbf{e_2} + z^2\mathbf{e_3}$. Determine the maximum value of $\left|\frac{d\mathbf{v}}{dr}\right|$ at the point (1,2,3).

1.126 Let $f = f(\phi, \psi)$ be a scalar function where ϕ and ψ are scalar fields. Show that the following relation holds:

$$\nabla f = \frac{\partial f}{\partial \phi}\nabla\phi + \frac{\partial f}{\partial \psi}\nabla\psi$$

1.127 Let Φ be a scalar field given by $\Phi = e^{az^3}\sin(bx)\cosh(cy)$ where a, b, and c are scalars. Determine the vector $\mathbf{v} = -\nabla\Phi$.

1.128 Find a vector normal to the surface $xy - z = 1$ at the point (2, 1, 1).

1.129 Let **r** be the position vector for a general point (x, y, z) in three-dimensional space. Show that:

(a) div **r** = 3

(b) **curl r** = **0**

(c) div $(r^{-3}\mathbf{r}) = 0$

where r is the length of the vector **r**.

1.130 Let **r** be the radius vector of a typical point in a field and r be the length of **r**. Show that:

(a) div $(r^n \mathbf{r}) = (n + 3) r^n$

(b) **curl** $(r^n \mathbf{r}) = \mathbf{0}$

(c) $\nabla (\nabla (r^n)) = n (n + 1) r^{n-2}$

1.131 Let Φ be a general scalar field and let **T** be a general tensor field. Show that the following relation holds:

$$\mathbf{div} \, (\Phi \mathbf{T}) = \mathbf{T} \, (\nabla \Phi) + \Phi \mathbf{div} \mathbf{T}$$

1.132 Let **T** be a symmetric tensor and **S** be an antisymmetric tensor. Show that tr(**TS**) = 0.

1.133 Let **Q**(t) be an orthogonal tensor. Show that $\frac{d\mathbf{Q}}{dt} \mathbf{Q}^T$ is an antisymmetric tensor.

1.134 Use polar coordinates with $\mathbf{e_r}$ and $\mathbf{e_\theta}$ representing unit vectors in the coordinate directions of **r** and θ, respectively. We have the position vector **r** and $d\mathbf{r}$ given by:

$$\mathbf{r} = r\mathbf{e_r}$$
$$d\mathbf{r} = dr\mathbf{e_r} + rd\theta\mathbf{e_\theta}$$

where r and dr are the lengths of **r** and $d\mathbf{r}$, respectively, and the vectors $\mathbf{e_r}$ and $\mathbf{e_\theta}$ are given by the following two equations:

$$\mathbf{e_r} = \cos\theta\mathbf{e_1} + \sin\theta\mathbf{e_2}$$

$$\mathbf{e_\theta} = -\sin\theta\mathbf{e_1} + \cos\theta\mathbf{e_2}$$

Prove the following relations using polar coordinates:

(a)

$$\nabla f = \frac{\partial f}{\partial r}\mathbf{e_r} + \frac{1}{r}\frac{\partial f}{\partial \theta}\mathbf{e_\theta}$$

where f is a scalar valued function, i.e., $f \equiv f(r, \theta)$.

(b) $\nabla \mathbf{v}$ is represented by the following matrix:

$$\nabla \mathbf{v} \equiv \begin{bmatrix} \frac{\partial v_r}{\partial r} & \frac{1}{r}\left(\frac{\partial v_r}{\partial \theta} - v_\theta\right) \\ \frac{\partial v_\theta}{\partial r} & \frac{1}{r}\left(\frac{\partial v_\theta}{\partial \theta} + v_r\right) \end{bmatrix}$$

where $\mathbf{v}$ is a vector field, i.e. $\mathbf{v} = v_r\mathbf{e_r} + v_\theta\mathbf{e_\theta}$ and both v_r and v_θ are scalar functions, i.e. $v_r(r, \theta)$ and $v_\theta(r, \theta)$.

(c) $\operatorname{div} \mathbf{v} = \frac{\partial v_r}{\partial r} + \frac{1}{r}\left(\frac{\partial v_\theta}{\partial \theta} + v_r\right)$

(d) $\operatorname{\mathbf{curl}} \mathbf{v} = \left(\frac{\partial v_\theta}{\partial r} + \frac{v_\theta}{r} - \frac{1}{r}\frac{\partial v_r}{\partial \theta}\right)\mathbf{e_3}$

(e)

$$(\operatorname{\mathbf{div}}\mathbf{T})_r = \frac{\partial T_{rr}}{\partial r} + \frac{1}{r}\frac{\partial T_{r\theta}}{\partial \theta} + \frac{T_{rr} - T_{\theta\theta}}{r}$$

$$(\operatorname{\mathbf{div}}\mathbf{T})_\theta = \frac{\partial T_{\theta r}}{\partial r} + \frac{1}{r}\frac{\partial T_{\theta\theta}}{\partial \theta} + \frac{T_{r\theta} + T_{\theta r}}{r}$$

where $\mathbf{T}$ is a tensor field, represented by the matrix:

$$\mathbf{T} \equiv \begin{bmatrix} T_{rr} & T_{r\theta} \\ T_{\theta r} & T_{\theta\theta} \end{bmatrix}$$

1.9 Maple Tensor Commands

In this section we present the commands for many tensor operations appearing in sections 1.6, 1.7, and 1.8. Concerning section 1.7 on eigenvalues and eigenvectors of second-order tensor, we can determine the eigenvalues and the eigenvectors of their corresponding matrix representation. This can be performed using the Maple commands *eigenvals* and *eigenvects*. See section 1.3 for more details on these two commands.

Next, we present the three Maple commands *grad, diverge* and *curl* for the vector and tensor operations in section 1.8. The Maple command *grad* is used to determine the gradient vector of a scalar function. This is illustrated by the following example:

```
>  grad(x^2+y^2+z^2,[x,y,z]);
```
$$[2\,x,\, 2\,y,\, 2\,z]$$

```
>  sf:=f(x,y,z);
```
$$sf := \mathrm{f}(x,\, y,\, z)$$

```
>  grad(sf,[x,y,z]);
```
$$\left[\tfrac{\partial}{\partial x}\,\mathrm{f}(x,\, y,\, z),\; \tfrac{\partial}{\partial y}\,\mathrm{f}(x,\, y,\, z),\; \tfrac{\partial}{\partial z}\,\mathrm{f}(x,\, y,\, z)\right]$$

In performing the above commands, you must invoke the linear algebra package first using the command *with(linalg);*.

The Maple command *diverge* determines the divergence of vector functions. The following examples illustrate the use of this command:

```
>  diverge([x*cos(y),y*cos(x),z],[x,y,z]);
```
$$\cos(y) + \cos(x) + 1$$

```
>  vf:=[fx(x,y,z),fy(x,y,z),fz(x,y,z)];
```
$$vf := [\mathrm{fx}(x,\, y,\, z),\, \mathrm{fy}(x,\, y,\, z),\, \mathrm{fz}(x,\, y,\, z)]$$

```
>  diverge(vf,[x,y,z]);
```
$$(\tfrac{\partial}{\partial x}\,\mathrm{fx}(x,\, y,\, z)) + (\tfrac{\partial}{\partial y}\,\mathrm{fy}(x,\, y,\, z)) + (\tfrac{\partial}{\partial z}\,\mathrm{fz}(x,\, y,\, z))$$

The Maple command *curl* is used to calculate the curl of a vector function. The following examples illustrate the use of this command:

```
>  curl([x*cos(y),y*cos(x),z],[x,y,z]);
```
$$[0,\, 0,\, -y\sin(x) + x\sin(y)]$$

```
>  vf:=[fx(x,y,z),fy(x,y,z),fz(x,y,z)];
```
$$vf := [\mathrm{fx}(x,\, y,\, z),\, \mathrm{fy}(x,\, y,\, z),\, \mathrm{fz}(x,\, y,\, z)]$$

```
>  curl(vf,[x,y,z]);
```

$$\left[(\tfrac{\partial}{\partial y}\,\mathrm{fz}(x,\, y,\, z)) - (\tfrac{\partial}{\partial z}\,\mathrm{fy}(x,\, y,\, z)),\; (\tfrac{\partial}{\partial z}\,\mathrm{fx}(x,\, y,\, z)) - (\tfrac{\partial}{\partial x}\,\mathrm{fz}(x,\, y,\, z)),\right.$$

$$\left.(\tfrac{\partial}{\partial x}\,\mathrm{fy}(x,\, y,\, z)) - (\tfrac{\partial}{\partial y}\,\mathrm{fx}(x,\, y,\, z))\right]$$

Finally, it should be noted that all three commands *grad, diverge,* and *curl* can be used for non-Cartesian coordinates, like cylindrical coordinates. This will be explored further in Problems 1.144 and 1.145.

Problems

1.135 Let Φ be a scalar function given by $\Phi(x, y, z) = x \cos y + y \cos z + z \cos x$. Determine the gradient vector $\nabla \Phi$ using Maple.

1.136 Solve Example 1.19 again using Maple.

1.137 Solve Problem 1.114 again using Maple.

1.138 Let $\mathbf{u}$ be a vector field given by $\mathbf{u} = \left(x^2 - y^2\right) \mathbf{e_1} + \left(z^2 - x^2\right) \mathbf{e_2} + \left(y^2 - z^2\right) \mathbf{e_3}$. Determine the three quantities $\nabla \mathbf{u}$, div $\mathbf{u}$ and **curlu** using Maple.

1.139 Solve Example 1.20 again using Maple.

1.140 Solve Example 1.22 again using Maple.

1.141 Solve Problem 1.115 again using Maple.

1.142 Solve Problem 1.127 again using Maple.

1.143 Solve Problem 1.128 again using Maple.

1.144 Let $\mathbf{v}$ be a vector field given as follows in cylindrical coordinates, $\mathbf{v} = r \cos \theta \mathbf{e_1} + r \sin \theta \mathbf{e_2} + r \mathbf{e_3}$, where r and θ are scalars. Determine $\nabla \mathbf{v}$, div $\mathbf{v}$ and **curlv** using Maple.

1.145 Solve Problem 1.134 again using Maple.

2

Elasticity Theory

In this chapter we present a summary of the theory of elasticity including the concepts of force, deformation, stress, and strain. This is achieved using both the material description and the spatial description. The theory of elasticity is reviewed in this chapter based on the book of Lai et al.(1984).

2.1 Motion of a Continuum

Consider a body occupying a certain region of space at time $t = t_o$. Let $\mathbf{X}$ denote the position vector of a particle in the body measured from an origin point O (see Figure 2.1). Then, the motion of every particle is described by the equation

$$\mathbf{x} = \mathbf{x}(\mathbf{X}, t)$$

$$\mathbf{X} = \mathbf{x}(\mathbf{X}, t_o) \tag{2.1}$$

where $\mathbf{X}$ denotes the position of the particle at $t = t_o$.

During the motion of a continuum, we describe it using either one of two approaches as follows:

1. By following the particles, i.e., we express the variables as functions of the particle material coordinates $\mathbf{X}$ and time t. This approach is known as the *material description* or the *Lagrangian description*.

2. By observing the changes at fixed locations, i.e., we express the variables as functions of fixed position x and time t. This approach is known as the *spatial description* or the *Eulerian description*.

Consequently, the coordinates $\mathbf{X}$ are called the *material coordinates* while the coordinates x are known as the *spatial coordinates*. Equation 2.1 relates these two types of coordinates and can be written explicitly as follows:

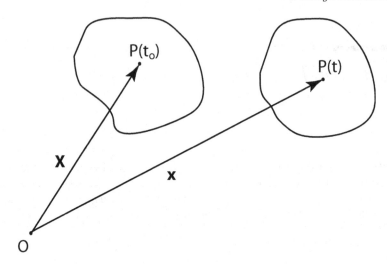

FIGURE 2.1
A particle in a body at times t_0 and t with position vectors $\mathbf{X}$ and $\mathbf{x}$,
respectively.(Lai et al., 1984)

$$x_1 = x_1(X_1, X_2, X_3, t)$$
$$x_2 = x_2(X_1, X_2, X_3, t)$$
$$x_3 = x_3(X_1, X_2, X_3, t) \tag{2.2}$$

The *material description* is defined as the time rate of change of a quantity of a material particle. Let D/Dt denote the material derivative. We next explore the material derivative using both the material description and the spatial description of motion.

1. Using the material description of a quantity like $\phi(\mathbf{X}, t)$, then the material derivative $D\phi/Dt$ is given by:

$$\frac{D\phi}{Dt} = \left. \frac{\partial\phi}{\partial t} \right|_{\mathbf{x} \text{ is fixed}} \tag{2.3}$$

2. Using the spatial description of a quantity like $\phi(\mathbf{x}, t)$, then the material derivative $D\phi/Dt$ is given by:

$$\frac{D\phi}{Dt} = \left. \frac{\partial\phi}{\partial t} \right|_{\mathbf{X} \text{ is fixed}} = \frac{\partial\phi}{\partial x_1}\frac{\partial x_1}{\partial t} + \frac{\partial\phi}{\partial x_2}\frac{\partial x_2}{\partial t} + \frac{\partial\phi}{\partial x_3}\frac{\partial x_3}{\partial t} + \left. \frac{\partial\phi}{\partial t} \right|_{\mathbf{x} \text{ is fixed}} \tag{2.4}$$

Note in the above equation that $\partial x_i / \partial t = v_i$ where v_i is the component of the velocity of the particle along x_i. Therefore, equation (2.4) can be written in compact form as follows:

$$\frac{D\phi}{Dt} = \frac{\partial \phi}{\partial t} + \mathbf{v} \cdot \nabla \phi \tag{2.5}$$

where $\nabla \phi$ is the gradient vector of the quantity ϕ. Note that equation (2.5) is valid for all coordinate systems, while equation (2.4) is valid only for rectangular Cartesian coordinates.

The acceleration of a particle is the material derivative of the velocity of the particle. The velocity $\mathbf{v}$ and acceleration $\mathbf{a}$, at time t, of a particle $\mathbf{X}$, are defined as follows:

$$\mathbf{v} = \frac{D\mathbf{x}}{Dt} \tag{2.6a}$$

$$\mathbf{a} = \frac{D\mathbf{v}}{Dt} \tag{2.6b}$$

Next, we explore the acceleration of a particle using both the material description and the spatial description of motion.

1. Using the material description of velocity $\mathbf{v}(\mathbf{X}, t)$, it is straight forward to compute the acceleration $\mathbf{a}(\mathbf{X},t)$ as follows:

$$\mathbf{a} = \frac{D\mathbf{v}}{Dt} = \left. \frac{\partial \mathbf{v}}{\partial t} \right|_{\mathbf{X} \text{ is fixed}} \tag{2.7}$$

2. Using the spatial description of velocity $\mathbf{v}(\mathbf{x}, t)$, we need to compute the acceleration $\mathbf{a}(\mathbf{x}, t)$ as follows:

$$\mathbf{a} = \frac{D\mathbf{v}}{Dt} = \frac{Dv_1}{Dt}\mathbf{e}_1 + \frac{Dv_2}{Dt}\mathbf{e}_2 + \frac{Dv_3}{Dt}\mathbf{e}_3 \tag{2.8a}$$

$$\frac{Dv_i}{Dt} = \frac{\partial v_i}{\partial t} + v_j \frac{\partial v_i}{\partial x_j} \tag{2.8b}$$

Therefore, we obtain:

$$a_i = \frac{\partial v_i}{\partial t} + v_j \frac{\partial v_i}{\partial x_j} \tag{2.9}$$

or equivalently we have:

$$\mathbf{a} = \frac{\partial \mathbf{v}}{\partial t} + (\nabla \mathbf{v})\mathbf{v} \tag{2.10}$$

Example 2.1

Given the motion of a body as follows:

$$x_1 = X_1$$
$$x_2 = X_2 + \alpha t X_1$$
$$x_3 = X_3$$

where α is constant.

Let the temperature field be given by the spatial description as $T = \beta x_1 + \gamma x_2$ where β and γ are constants.

    a. Determine the material description of temperature $T(X_1, X_2, X_3, t)$.

    b. Determine the velocity and rate of change of temperature using both the material description and the spatial description.

Solution

    a.

$$T(X_1, X_2, X_3, t) = \beta x_1 + \gamma x_2$$
$$= \beta X_1 + \gamma(X_2 + \alpha t X_1)$$
$$T = (\beta + \alpha \gamma t) X_1 + \gamma X_2$$

    b.

$$v_i = \left. \frac{\partial x_i}{\partial t} \right|_{\mathbf{X} \text{ is fixed}}$$

$$\text{Therefore,} \quad v_1 = \frac{\partial x_1}{\partial t} = 0$$

$$v_2 = \frac{\partial x_2}{\partial t} = \alpha X_1$$

$$v_3 = \frac{\partial x_3}{\partial t} = 0$$

The above equations represent the material description of velocity. The following is the spatial description:

$$v_1 = 0$$
$$v_2 = \alpha x_1$$
$$v_3 = 0$$

The rate of change of temperature $\partial\theta/\partial t$ is given as follows in both descriptions:

$$\frac{\partial\theta}{\partial t} = \alpha\gamma X_1 \quad \text{using the material description}$$

$$\frac{\partial\theta}{\partial t} = \alpha\gamma x_1 \quad \text{using the spatial description}$$

Example 2.2

Given the following motion

$$x_1 = X_1$$
$$x_2 = X_2$$
$$x_3 = X_3 + \alpha t$$

etermine the velocity and acceleration of a particle using both the material description and spatial description where α is a constant.

Solution

In the material description, we have:

$$v_1 = 0$$
$$v_2 = 0$$
$$v_3 = \alpha$$
$$a_1 = 0$$
$$a_2 = 0$$
$$a_3 = 0$$

sing the spatial description, we have:

$$v_1 = 0$$
$$v_2 = 0$$
$$v_3 = \alpha$$
$$a_1 = 0$$
$$a_2 = 0$$
$$a_3 = 0$$

Problems

2.1. Given the following motion using the material description:

$$x_1 = X_1 + \beta X_1$$
$$x_2 = X_2 + \alpha X_2$$
$$x_3 = X_3$$

where α and β are constants. Write the motion using the spatial description.

2.2. Describe the motion given in problem 2.1 geometrically.

2.3. Consider the motion given in problem 2.1.

(a) Determine the velocity field using the material description.

(b) Determine the velocity field using the spatial description.

(c) Determine the acceleration field using the material description.

(d) Determine the acceleration field using the spatial description.

2.4. Consider the motion given in Problem 2.1. Let ϕ be a scalar field given by $\phi = x_1 - x_2 + 2x_3$.

(a) Determine the material description of ϕ.

(b) Determine the rate of change of ϕ using the material description.

(c) Determine the rate of change of ϕ using the spatial description.

2.5. Consider a velocity field given by:

$$v_1 = \frac{x_1}{1-t}$$

$$v_2 = \frac{x_2}{1-t}$$

$$v_3 = \frac{x_3}{1-t}$$

Determine the acceleration field.

2.6. Consider the velocity field of problem 2.5. Determine the motion $\mathbf{x} = \mathbf{x}(\mathbf{X}, t)$.

2.7. Consider the following motion:

$$x_1 = X_1 - \alpha_1 t$$
$$x_2 = X_2 - \alpha_2 t$$
$$x_3 = X_3 - \alpha_3 t$$

where α_1, α_2 and α_3 are constants.

(a) Determine the motion using the spatial description.

(b) Determine the velocity field using both the material description and the spatial description.

(c) Determine the acceleration field using both the material description and the spatial description.

2.8. Consider the motion of Problem 2.7. Let ϕ be a scalar field given by $\phi = x_1 x_2 x_3$.

(a) Determine the material description of ϕ.

(b) Determine $\frac{D\phi}{Dt}$ using both the material description and the spatial description.

2.2   Deformation and Strain

Let a body deform from an initial configuration at t_o to a final configuration at t (see Figure 2.2). Let P and Q be two points separated by an infinitesimal distance $d\mathbf{X}$ initially and $d\mathbf{x}$ finally. Point P undergoes a displacement $\mathbf{u}$ such that:

$$\mathbf{x} = \mathbf{X} + \mathbf{u}(\mathbf{X}, t) \tag{2.11}$$

The neighboring point Q has a displacement $\mathbf{u}(\mathbf{X} + d\mathbf{X}, t)$ according to the equation:

$$d\mathbf{x} = d\mathbf{X} + \mathbf{u}(\mathbf{X} + d\mathbf{X}, t) - \mathbf{u}(\mathbf{X}, t) \tag{2.12}$$

The above equation is clear from the geometry shown in Figure 2.2. Equation (2.12) can be rewritten as follows:

$$d\mathbf{x} = d\mathbf{X} + (\nabla \mathbf{u}) d\mathbf{X} \tag{2.13}$$

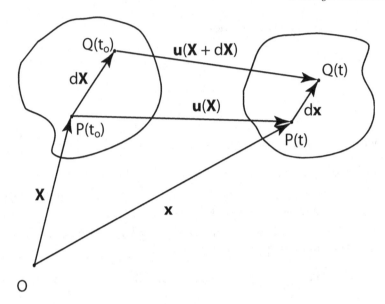

FIGURE 2.2
Deformation of a body between t_0 and t.(Lai et al., 1984)

where $\nabla \mathbf{u}$ is a second-order tensor called the displacement gradient. The tensor $\nabla \mathbf{u}$ has the following matrix representation:

$$\nabla \mathbf{u} \equiv \begin{bmatrix} \frac{\partial u_1}{\partial X_1} & \frac{\partial u_1}{\partial X_2} & \frac{\partial u_1}{\partial X_3} \\ \frac{\partial u_2}{\partial X_1} & \frac{\partial u_2}{\partial X_2} & \frac{\partial u_2}{\partial X_3} \\ \frac{\partial u_3}{\partial X_1} & \frac{\partial u_3}{\partial X_2} & \frac{\partial u_3}{\partial X_3} \end{bmatrix} \tag{2.14}$$

It is clear from equation (2.13) that if $d\mathbf{x} = d\mathbf{X}$, then $\nabla \mathbf{u} = \mathbf{0}$. This special case corresponds to a rigid body motion.

Consider now two material vectors $d\mathbf{X}^1$ and $d\mathbf{X}^2$ at point P. After deformation, they become $d\mathbf{x}^1$ and $d\mathbf{x}^2$, respectively, such that (from equation (2.13)):

$$d\mathbf{x}^1 = d\mathbf{X}^1 + (\nabla \mathbf{u})(d\mathbf{X}^1) \tag{2.15a}$$

$$d\mathbf{x}^2 = d\mathbf{X}^2 + (\nabla \mathbf{u})(d\mathbf{X}^2) \tag{2.15b}$$

Consider now the dot product $d\mathbf{x}^1 \cdot d\mathbf{x}^2$:

$$d\mathbf{x}^1 \cdot d\mathbf{x}^2 = d\mathbf{X}^1 \cdot d\mathbf{X}^2 + d\mathbf{X}^1 \cdot (\nabla \mathbf{u})(d\mathbf{X}^2)$$
$$+ \; d\mathbf{X}^2 \cdot (\nabla \mathbf{u})(d\mathbf{X}^1) + [(\nabla \mathbf{u})(d\mathbf{X}^1)] \cdot [(\nabla \mathbf{u})(d\mathbf{X}^2)] \tag{2.16}$$

Simplifying equation (2.16) we obtain:

$$d\mathbf{x}^1 \cdot d\mathbf{x}^2 = d\mathbf{X}^1 \cdot d\mathbf{X}^2 + d\mathbf{X}^1 \cdot [(\nabla\mathbf{u}) + (\nabla\mathbf{u})^T + (\nabla\mathbf{u})^T(\nabla\mathbf{u})](d\mathbf{X}^2) \quad (2.17)$$

Define the second-order tensor $\mathbf{E}^*$ as follows:

$$\mathbf{E} = \frac{1}{2}[(\nabla\mathbf{u}) + (\nabla\mathbf{u})^T + (\nabla\mathbf{u})^T(\nabla\mathbf{u})] \quad (2.18)$$

Next we substitute equation (2.18) into equation (2.17) to obtain:

$$d\mathbf{x}^1 \cdot d\mathbf{x}^2 = d\mathbf{X}^1 \cdot d\mathbf{X}^2 + 2d\mathbf{X}^1 \cdot \mathbf{E}^*(d\mathbf{X}^2) \quad (2.19)$$

It is clear from equation (2.19) that if $\mathbf{E}^* = \mathbf{0}$, then $d\mathbf{x}^1 \cdot d\mathbf{x}^2 = d\mathbf{X}^1 \cdot d\mathbf{X}^2$, i.e., $\mathbf{E}^*$ characterizes the deformation of the neighborhood of the particle P. The tensor $\mathbf{E}^*$ is called the *Lagrangian Strain Tensor*.

Using equation (2.18), we can write the components of $\mathbf{E}^*$ with respect to rectangular Cartesian coordinates as follows:

$$E_{ij}^* = \frac{1}{2}\left(\frac{\partial u_i}{\partial X_j} + \frac{\partial u_j}{\partial X_i} + \frac{\partial u_k}{\partial X_i}\frac{\partial u_k}{\partial X_j}\right) \quad (2.20)$$

Equations (2.18) and (2.20) apply to general cases of deformation involving large strains. For the special case of small strains and small deformations, the term $(\nabla\mathbf{u})^T(\nabla\mathbf{u})$ can be neglected. Therefore, equation (2.18) becomes:

$$\mathbf{E}^* \cong \frac{1}{2}[(\nabla\mathbf{u}) + (\nabla\mathbf{u})^T] = (\nabla\mathbf{u})^S \quad (2.21)$$

where $(\nabla\mathbf{u})^S$ is the symmetric part of $\nabla\mathbf{u}$.

Let $\mathbf{E}$ denote the *strain tensor* for small deformation, also called the *infinitesimal strain tensor*. Then equations (2.19) and (2.21) become:

$$d\mathbf{x}^1 \cdot d\mathbf{x}^2 = d\mathbf{X}^1 \cdot d\mathbf{X}^2 + 2d\mathbf{X}^1 \cdot \mathbf{E}(d\mathbf{X}^2) \quad (2.22)$$

$$\mathbf{E} = \frac{1}{2}[(\nabla\mathbf{u}) + (\nabla\mathbf{u})^T] \quad (2.23)$$

It should be noted that the anti-symmetric part of $\nabla\mathbf{u}$ is called the *infinitesimal rotation tensor* or just the *rotation tensor*, denoted $\mathbf{\Omega}$ and given by :

$$\mathbf{\Omega} = \frac{1}{2}\left[(\nabla\mathbf{u}) - (\nabla\mathbf{u})^T\right] \quad (2.24)$$

The angle and axis of the rotation is determined by the dual vector of $\mathbf{\Omega}$.

Next, we write the components of **E** explicitly as follows (see equation (2.20)):

$$E_{ij} = \frac{1}{2}\left(\frac{\partial u_i}{\partial X_j} + \frac{\partial u_j}{\partial X_i}\right) \tag{2.25}$$

The matrix representation of **E** is given as follows:

$$\mathbf{E} \equiv \begin{bmatrix} \frac{\partial u_1}{\partial X_1} & \frac{1}{2}\left(\frac{\partial u_1}{\partial X_2} + \frac{\partial u_2}{\partial X_1}\right) & \frac{1}{2}\left(\frac{\partial u_1}{\partial X_3} + \frac{\partial u_3}{\partial X_1}\right) \\ \frac{1}{2}\left(\frac{\partial u_1}{\partial X_2} + \frac{\partial u_2}{\partial X_1}\right) & \frac{\partial u_2}{\partial X_2} & \frac{1}{2}\left(\frac{\partial u_2}{\partial X_3} + \frac{\partial u_3}{\partial X_2}\right) \\ \frac{1}{2}\left(\frac{\partial u_1}{\partial X_3} + \frac{\partial u_3}{\partial X_1}\right) & \frac{1}{2}\left(\frac{\partial u_2}{\partial X_3} + \frac{\partial u_3}{\partial X_2}\right) & \frac{\partial u_3}{\partial X_3} \end{bmatrix} \tag{2.26}$$

Next, we will discuss the rate of deformation. Consider a material segment dx at a material point P located at $\mathbf{x}$ at time t. Using $\mathbf{x} = \mathbf{x}(X, t)$ we can obtain:

$$dx = \mathbf{x}(\mathbf{X} + d\mathbf{X}, t) - \mathbf{x}(\mathbf{X}, t) \tag{2.27}$$

Therefore, the material derivative $\frac{D}{Dt}(dx)$ is obtained as follows:

$$\frac{D}{Dt}(dx) = \mathbf{v}(\mathbf{X} + d\mathbf{X}, t) - \mathbf{v}(X, t) = (\nabla_{\mathbf{X}}\mathbf{v})(d\mathbf{X}) \tag{2.28}$$

where $(\nabla_{\mathbf{X}}\mathbf{v})$ is the gradient of **v** with respect to the material coordinate X. Equation (2.28) represents the material derivative $\frac{D}{Dt}(dx)$ using the material description. We write now $\frac{D}{Dt}(dx)$ using the spatial description as follows:

$$\frac{D}{Dt}(dx) = \mathbf{v}(\mathbf{x} + dx, t) - \mathbf{v}(\mathbf{x}, t) = (\nabla_{\mathbf{x}}\mathbf{v})(dx) \tag{2.29}$$

where $(\nabla_{\mathbf{x}}\mathbf{v})$ is the spatial gradient of velocity. We will use $\nabla\mathbf{v}$ instead of $\nabla_{\mathbf{x}}\mathbf{v}$ to denote this spatial gradient in the following derivations. The matrix representation of $\nabla\mathbf{v}$ is given as follows:

$$\nabla\mathbf{v} \equiv \begin{bmatrix} \frac{\partial v_1}{\partial x_1} & \frac{\partial v_1}{\partial x_2} & \frac{\partial v_1}{\partial x_3} \\ \frac{\partial v_2}{\partial x_1} & \frac{\partial v_2}{\partial x_2} & \frac{\partial v_2}{\partial x_3} \\ \frac{\partial v_3}{\partial x_1} & \frac{\partial v_3}{\partial x_2} & \frac{\partial v_3}{\partial x_3} \end{bmatrix} \tag{2.30}$$

In general, $\nabla\mathbf{v}$ can be decomposed into its symmetric and anti-symmetric parts as follows:

$$\nabla\mathbf{v} = \mathbf{D} + \mathbf{W} \tag{2.31}$$

where **D** is called the *rate of deformation tensor* given by:

$$D = \frac{(\nabla v) + (\nabla v)^T}{2} \qquad (2.32)$$

and **W** is called the *spin tensor* given by:

$$W = \frac{(\nabla v) - (\nabla v)^T}{2} \qquad (2.33)$$

The matrix representations of both **D** and **W** are given explicitly as follows based on equations (2.32) and (2.33):

$$D \equiv \begin{bmatrix} \frac{\partial v_1}{\partial x_1} & \frac{1}{2}\left(\frac{\partial v_1}{\partial x_2} + \frac{\partial v_2}{\partial x_1}\right) & \frac{1}{2}\left(\frac{\partial v_1}{\partial x_3} + \frac{\partial v_3}{\partial x_1}\right) \\ \frac{1}{2}\left(\frac{\partial v_1}{\partial x_2} + \frac{\partial v_2}{\partial x_1}\right) & \frac{\partial v_2}{\partial x_2} & \frac{1}{2}\left(\frac{\partial v_2}{\partial x_3} + \frac{\partial v_3}{\partial x_2}\right) \\ \frac{1}{2}\left(\frac{\partial v_1}{\partial x_3} + \frac{\partial v_3}{\partial x_1}\right) & \frac{1}{2}\left(\frac{\partial v_2}{\partial x_3} + \frac{\partial v_3}{\partial x_2}\right) & \frac{\partial v_3}{\partial x_3} \end{bmatrix} \qquad (2.34)$$

$$W \equiv \begin{bmatrix} 0 & \frac{1}{2}\left(\frac{\partial v_1}{\partial x_2} - \frac{\partial v_2}{\partial x_1}\right) & \frac{1}{2}\left(\frac{\partial v_1}{\partial x_3} - \frac{\partial v_3}{\partial x_1}\right) \\ -\frac{1}{2}\left(\frac{\partial v_1}{\partial x_2} - \frac{\partial v_2}{\partial x_1}\right) & 0 & \frac{1}{2}\left(\frac{\partial v_2}{\partial x_3} - \frac{\partial v_3}{\partial x_2}\right) \\ -\frac{1}{2}\left(\frac{\partial v_1}{\partial x_3} - \frac{\partial v_3}{\partial x_1}\right) & -\frac{1}{2}\left(\frac{\partial v_2}{\partial x_3} - \frac{\partial v_3}{\partial x_2}\right) & 0 \end{bmatrix} \qquad (2.35)$$

Example 2.3

Given the following displacement field:

$$u_1 = 0.001X_2^2$$
$$u_2 = 0$$
$$u_3 = 0$$

Determine the infinitesimal strain tensor **E**.

Solution

First, determine the matrix representation of ∇u as follows:

$$\nabla u \equiv \begin{bmatrix} 0 & 0.002X_2 & 0 \\ 0 & 0 & 0 \\ 0 & 0 & 0 \end{bmatrix}$$

The infinitesimal strain tensor is then obtained as the symmetric part of ∇u:

$$E = (\nabla u)^S \equiv \begin{bmatrix} 0 & 0.001X_2 & 0 \\ 0.001X_2 & 0 & 0 \\ 0 & 0 & 0 \end{bmatrix}$$

Example 2.4

Given the following velocity field:

$$v_1 = \alpha x_2$$
$$v_2 = 0$$
$$v_3 = 0$$

Determine the rate of deformation tensor $\mathbf{D}$ and the spin tensor $\mathbf{W}$.

Solution

First, we determine the matrix representation of $\nabla \mathbf{v}$ as follows:

$$\nabla \mathbf{v} \equiv \begin{bmatrix} 0 & \alpha & 0 \\ 0 & 0 & 0 \\ 0 & 0 & 0 \end{bmatrix}$$

Then, the tensors $\mathbf{D}$ and $\mathbf{W}$ are determined as the symmetric and anti-symmetric parts of $\nabla \mathbf{v}$, respectively, as follows:

$$\mathbf{D} = (\nabla \mathbf{v})^S = \begin{bmatrix} 0 & \frac{\alpha}{2} & 0 \\ \frac{\alpha}{2} & 0 & 0 \\ 0 & 0 & 0 \end{bmatrix}$$

$$\mathbf{W} = (\nabla \mathbf{v})^A = \begin{bmatrix} 0 & \frac{\alpha}{2} & 0 \\ -\frac{\alpha}{2} & 0 & 0 \\ 0 & 0 & 0 \end{bmatrix}$$

Problems

2.9. Given the following displacement field:

$$u_1 = k_1 X_2^2$$
$$u_2 = k_2 X_1^2$$
$$u_3 = 0$$

where k_1 and k_2 are constants. Determine the infinitesimal strain tensor $\mathbf{E}$.

2.10. Consider the displacement field given in Problem 2.9. Using the strain tensor $\mathbf{E}$, determine the unit elongation for the material elements $dX^1 = dX_1 \mathbf{e}_1$ and $dX^2 = dX_2 \mathbf{e}_2$ which were at the point $(1,1,0)$ of a unit square. Also, determine the decrease in angle between these two elements.

2.11. Given the following displacement field:

$$u_1 = \alpha(X_1^2 + \beta X_2^2)$$
$$u_2 = \alpha(X_1^2 - \beta X_2^2)$$
$$u_3 = 0$$

where $\alpha = 1 \times 10^{-4}$ and $\beta = 2$.

(a) Determine the unit elongations and the change of angle for two material elements $dX^1 = dX_1 e_1$ and $dX^2 = dX_2 e_2$ that emanate from a particle designated by $X = e_1 + e_2$.

(b) Determine the deformed position of these two elements dX^1 and dX^2.

2.12. Consider the following displacement field:

$$u_1 = 0$$
$$u_2 = \alpha X_2$$
$$u_3 = 0$$

where $\alpha = 1 \times 10^{-3}$. Let this displacement field act on a unit cube, with its edges parallel to the coordinate axes. Determine the increase in length of the diagonal AB (see Figure 2.3).

2.13. Given the following velocity field:

$$v_1 = \alpha_1(x_1 + x_2)$$
$$v_2 = \alpha_2(x_1 - x_2)$$
$$v_3 = 0$$

where α_1 and α_2 are scalars. Determine the rate of deformation tensor and spin tensor.

2.14. Consider the velocity field given in Problem 2.13.

(a) Determine the rate of extension of the material elements:

$$dx^1 = (ds_1)e_1$$
$$dx^2 = (ds_2)e_2$$
$$dx = dl(2e_1 + e_2)$$

(b) Determine the maximum and minimum rates of extension.

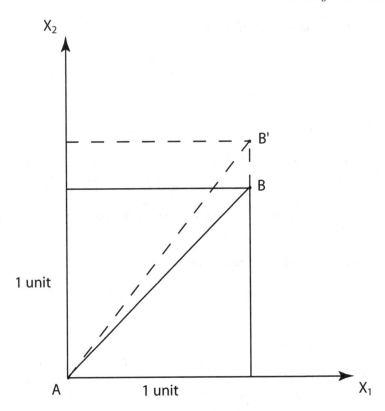

FIGURE 2.3
Unit cube subjected to a displacement field.

2.15. Given the following strain tensor with reference to an x_1 - x_2 - x_3 - coordinate system:

$$\mathbf{E} \equiv \begin{bmatrix} 1 & 2 & 0 \\ 2 & 5 & -1 \\ 0 & -1 & 3 \end{bmatrix} \times 10^{-4}$$

(a) Determine the unit elongation in the direction $\mathbf{e}_1 + \mathbf{e}_2 + \mathbf{e}_3$.

(b) What is the change of angle between two perpendicular lines (in the undeformed state) emanating from the point and in the direction of $2\mathbf{e}_1 + 2\mathbf{e}_2 + 2\mathbf{e}_3$ and $3\mathbf{e}_1 - 6\mathbf{e}_3$.

2.16. Consider the state of strain given in Problem 2.16. Determine the scalar invariants of the state of strain.

2.17. Show that the following matrix

$$\begin{bmatrix} 3 & 0 & 0 \\ 0 & 2 & 0 \\ 0 & 0 & 5 \end{bmatrix} \times 10^{-4}$$

cannot represent the same state of strain given in Problem 2.16.

2.18. Consider a material element $\mathbf{dx} = (ds)\mathbf{n}$.

(a) Show that $(D/Dt)\mathbf{n} = \mathbf{Dn} + \mathbf{Wn} - (\mathbf{n} \cdot \mathbf{D(n)})\mathbf{n}$.

(b) Show that if $\mathbf{n}$ is an eigenvector of $\mathbf{D}$ that $(D/Dt)\mathbf{n} = \mathbf{Wn} = \boldsymbol{\omega} \times \mathbf{n}$ where $\boldsymbol{\omega}$ is the dual vector of $\mathbf{W}$.

2.19. Prove the six *equations of compatibility* of strain components given below:

$$\frac{\partial^2 E_{11}}{\partial X_2^2} + \frac{\partial^2 E_{22}}{\partial X_1^2} = 2\frac{\partial^2 E_{12}}{\partial X_1 \partial X_2} \tag{2.36a}$$

$$\frac{\partial^2 E_{22}}{\partial X_3^2} + \frac{\partial^2 E_{33}}{\partial X_2^2} = 2\frac{\partial^2 E_{23}}{\partial X_2 \partial X_3} \tag{2.36b}$$

$$\frac{\partial^2 E_{33}}{\partial X_1^2} + \frac{\partial^2 E_{11}}{\partial X_3^2} = 2\frac{\partial^2 E_{31}}{\partial X_3 \partial X_1} \tag{2.36c}$$

$$\frac{\partial^2 E_{11}}{\partial X_2 \partial X_3} = \frac{\partial}{\partial X_1}\left(-\frac{\partial E_{33}}{\partial X_1} + \frac{\partial E_{31}}{\partial X_2} + \frac{\partial E_{12}}{\partial X_3}\right) \tag{2.36d}$$

$$\frac{\partial^2 E_{22}}{\partial X_3 \partial X_1} = \frac{\partial}{\partial X_2}\left(-\frac{\partial E_{31}}{\partial X_2} + \frac{\partial E_{12}}{\partial X_3} + \frac{\partial E_{23}}{\partial X_1}\right) \tag{2.36e}$$

$$\frac{\partial^2 E_{33}}{\partial X_1 \partial X_2} = \frac{\partial}{\partial X_1}\left(-\frac{\partial E_{12}}{\partial X_3} + \frac{\partial E_{23}}{\partial X_1} + \frac{\partial E_{31}}{\partial X_2}\right) \tag{2.36f}$$

2.20. Check whether or not the following state of strain satisfies the compatibility conditions given in Problem 2.19.

$$\mathbf{E} \equiv \begin{bmatrix} X_1^2 & X_1 + X_2 & X_1 + X_3 \\ X_1 + X_2 & X_2^2 & X_2 + X_3 \\ X_1 + X_3 & X_2 + X_3 & X_3^2 \end{bmatrix}$$

2.21. Consider the following displacement field:

$$u_1 = \sin X_1$$
$$u_2 = \cos X_2$$
$$u_3 = \tan X_3$$

Does the above displacement field correspond to a strain field that satisfies the compatibility conditions given in Problem 2.19.

2.22. Give an example of a displacement field with a corresponding strain field that satisfies the compatibility equations given in Problem 2.19.

2.23. Consider the following strain field:

$$\mathbf{E} \equiv \begin{bmatrix} \frac{1}{\alpha}f(X_2, X_3) & 0 & 0 \\ 0 & -\frac{\nu}{\alpha}f(X_2, X_3) & 0 \\ 0 & 0 & -\frac{\nu}{\alpha}f(X_2, X_3) \end{bmatrix}$$

where ν and α are constants. What restrictions must $f(X_2, X_3)$ satisfy so that the above strain field $\mathbf{E}$ satisfies the compatibility equations given in Problem 2.19.

2.24. Write general expressions for the scalar invariants I_1, I_2, and I_3 of the strain tensor $\mathbf{E}$.

2.25. Define the *principal strain* and write a general expression for the principal strain matrix.

2.26. We define *dilatation*, denoted by e, as the unit volume change. Show that:

$$e = E_{ii} = E_{11} + E_{22} + E_{33} = \operatorname{div}\mathbf{u}$$

where we have assumed small deformation.

2.27. Derive the equations of compatibility for the rate of deformation components D_{ij} in a similar way to those given in Problem 2.19.

2.28. Consider the rate of deformation tensor $\mathbf{D}$ obtained after solving Problem 2.13. Does this tensor satisfy the compatibility conditions derived in Problem 2.27.

2.29. Give an example of a velocity field with a corresponding rate of deformation tensor that satisfies the compatibility equations derived in Problem 2.27.

2.30. Show that the acceleration components in cylindrical coordinates a_r, a_θ, and a_z, are given as follows:

$$a_r = \frac{\partial v_r}{\partial t} + v_r \frac{\partial v_r}{\partial r} + \frac{v_\theta}{r} \frac{\partial v_r}{\partial \theta} + v_z \frac{\partial v_r}{\partial z} - \frac{v_\theta^2}{r}$$

$$a_\theta = \frac{\partial v_\theta}{\partial t} + v_r \frac{\partial v_\theta}{\partial r} + \frac{v_\theta}{r} \frac{\partial v_\theta}{\partial \theta} + v_z \frac{\partial v_\theta}{\partial z} + \frac{v_r v_\theta}{r}$$

$$a_z = \frac{\partial v_z}{\partial t} + v_r \frac{\partial v_z}{\partial r} + \frac{v_\theta}{r} \frac{\partial v_z}{\partial \theta} + v_z \frac{\partial v_z}{\partial z}$$

2.31. Let u_r, u_θ, and u_z be the displacement components in cylindrical coordinates. Show that the components of the strain tensor **E** in cylindrical coordinates are given by:

$$E_{rr} = \frac{\partial u_r}{\partial r}$$

$$E_{\theta\theta} = \frac{1}{r} \frac{\partial u_\theta}{\partial \theta} + \frac{u_r}{r}$$

$$E_{zz} = \frac{\partial u_z}{\partial z}$$

$$E_{r\theta} = \frac{1}{2} \left(\frac{1}{r} \frac{\partial u_r}{\partial \theta} - \frac{\partial u_\theta}{r} + \frac{\partial u_\theta}{\partial r} \right)$$

$$E_{\theta r} = E_{r\theta}$$

$$E_{rz} = \frac{1}{2} \left(\frac{\partial u_r}{\partial z} + \frac{\partial u_z}{\partial r} \right)$$

$$E_{zr} = E_{rz}$$

$$E_{\theta z} = \frac{1}{2} \left(\frac{\partial u_\theta}{\partial z} + \frac{1}{r} \frac{\partial u_z}{\partial \theta} \right)$$

$$E_{z\theta} = E_{\theta z}$$

2.32. Let v_r, v_θ, and v_z be the velocity components in cylindrical coordinates. Show that the components of both the rate of deformation tensor **D** and the spin tensor **W**, in cylindrical coordinates, are given by:

$$D_{rr} = \frac{\partial v_r}{\partial r}$$

$$D_{\theta\theta} = \frac{1}{r}\frac{\partial v_\theta}{\partial\theta} + \frac{v_r}{r}$$

$$D_{zz} = \frac{\partial v_z}{\partial z}$$

$$D_{r\theta} = \frac{1}{2}\left(\frac{1}{r}\frac{\partial v_r}{\partial\theta} - \frac{\partial v_\theta}{r} + \frac{\partial v_\theta}{\partial r}\right)$$

$$D_{\theta r} = D_{r\theta}$$

$$D_{rz} = \frac{1}{2}\left(\frac{\partial v_r}{\partial z} + \frac{\partial v_z}{\partial r}\right)$$

$$D_{zr} = D_{rz}$$

$$D_{\theta z} = \frac{1}{2}\left(\frac{\partial v_\theta}{\partial z} + \frac{1}{r}\frac{\partial v_z}{\partial\theta}\right) \text{ , symmetric.}$$

$$D_{z\theta} = D_{\theta z}$$

$$W_{rr} = 0$$

$$W_{\theta\theta} = 0$$

$$W_{zz} = 0$$

$$W_{r\theta} = \frac{1}{2}\left(\frac{1}{r}\frac{\partial v_r}{\partial\theta} - \frac{\partial v_\theta}{r} + \frac{\partial v_\theta}{\partial r}\right)$$

$$W_{\theta r} = -W_{r\theta}$$

$$W_{rz} = \frac{1}{2}\left(\frac{\partial v_r}{\partial z} - \frac{\partial v_z}{\partial r}\right)$$

$$W_{zr} = -W_{rz}$$

$$W_{\theta z} = \frac{1}{2}\left(\frac{\partial v_\theta}{\partial z} - \frac{1}{r}\frac{\partial v_z}{\partial\theta}\right)$$

$$W_{z\theta} = -W_{\theta z}$$

2.3   Stress

In this section we will describe the forces inside a body in terms of the stress vector and the stress tensor. In addition, the equations of the equilibrium will be formulated in terms of the stress tensor. Finally, principal stresses and principal directions will be discussed.

Consider a body subjected to external forces as shown in Figure 2.4. Let **n** be the unit normal at an arbitrary point P where a plane S passes

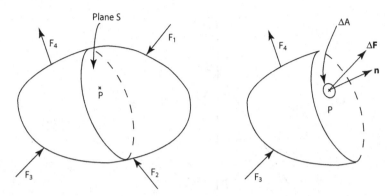

FIGURE 2.4

A sectional plane S is taken at an arbitrary point P in the body.(Lai et al., 1984)

as shown in the figure. Let ΔF be the resultant force acting on a small area ΔA around point P. The stress vector t_n at point P on the plane S is defined as follows:

$$t_n = \lim_{\Delta A \to 0} \frac{\Delta F}{\Delta A} \qquad (2.37)$$

where t_n points outward away from the surface S. Replacing the plane S and replacing the small area ΔA by a small surface area ΔS (on the surface), the stress vector at point P on the surface is then defined as follows:

$$t_n = \lim_{\Delta S \to 0} \frac{\Delta F}{\Delta S} \qquad (2.38)$$

It should be noted that $t \equiv t\,(x, t, n)$ which means that the stress vector at any given place x and time t has a common value on all parts of material having a common tangent plane at P and lying on the same side of it. This is known as the *Cauchy stress principle*.

Let n be the normal vector to a plane and let T be a transformation such that the stress vector on the plane is given by

$$t_n = T(n) \qquad (2.39)$$

We will show next that T is a linear transformation, i.e., a second-order tensor.

Consider a small tetrahedron isolated from the body with the point P as one of its vertices, as shown in Figure 2.5. It should be noted that the size (or volume) of the tetrahedron approaches zero so that the inclined plane passes through point P.

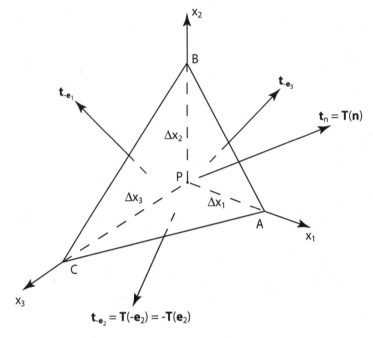

FIGURE 2.5
Stress tetrahedron.(Lai et al., 1984)

The stress vector $\mathbf{t}_{-\mathbf{e}_1}$ on the face PAB, whose outward normal is in the direction of $-\mathbf{e}_1$, is given by (based on equation (2.39)):

$$\mathbf{t}_{-\mathbf{e}_1} = -\mathbf{t}_{\mathbf{e}_1} = -\mathbf{T}(\mathbf{e}_1) \qquad (2.40a)$$

Similarly, the stress vectors $\mathbf{t}_{-\mathbf{e}_2}$ and $\mathbf{t}_{-\mathbf{e}_3}$ on the faces PBC and PAC, respectively, are given by:

$$\mathbf{t}_{-\mathbf{e}_2} = -\mathbf{T}(\mathbf{e}_2) \qquad (2.40b)$$

$$\mathbf{t}_{-\mathbf{e}_3} = -\mathbf{T}(\mathbf{e}_3) \qquad (2.40c)$$

Let ΔA_1, ΔA_2, and ΔA_3 be the areas of faces PAB, PBC, and PAC, respectively. Using Newton's second law, we can write the following:

$$\sum \mathbf{F} = \mathbf{t}_{-\mathbf{e}_1}(\Delta A_1) + \mathbf{t}_{-\mathbf{e}_2}(\Delta A_2) + \mathbf{t}_{-\mathbf{e}_3}(\Delta A_3) + \mathbf{t}_\mathbf{n}(\Delta A_\mathbf{n}) = m\mathbf{a} \quad (2.41)$$

where ΔA_n is the area of the inclined face ABC. When the volume of the tetrahedron approaches zero, the right-hand-side of the equation (2.41) vanishes, i.e. $m\mathbf{a} = 0$. Therefore, we obtain:

$$\mathbf{t}_{-\mathbf{e}_1}(\Delta A_1) + \mathbf{t}_{-\mathbf{e}_2}(\Delta A_2) + \mathbf{t}_{-\mathbf{e}_3}(\Delta A_3) + \mathbf{t}_{\mathbf{n}}(\Delta A_n) = 0 \qquad (2.42)$$

Let the unit vector $\mathbf{n}$ on the inclined plane ABC given by:

$$\mathbf{n} = n_1\mathbf{e}_1 + n_2\mathbf{e}_2 + n_3\mathbf{e}_3 \qquad (2.43)$$

We now can write the areas ΔA_1, ΔA_2, and ΔA_3 in terms of the inclined area ΔA_n, as follows:

$$\Delta A_1 = n_1(\Delta A_n) \qquad (2.44a)$$

$$\Delta A_2 = n_2(\Delta A_n) \qquad (2.44b)$$

$$\Delta A_3 = n_3(\Delta A_n) \qquad (2.44c)$$

Substituting equations (2.40) and (2.44) into equation (2.42) and simplifying, we obtain:

$$\mathbf{T}(\mathbf{n}) = n_1\mathbf{T}(\mathbf{e}_1) + n_2\mathbf{T}(\mathbf{e}_2) + n_3\mathbf{T}(\mathbf{e}_3) \qquad (2.45)$$

Combining equations (2.43) and (2.45) we can deduce that $\mathbf{T}$ is a linear transformation. $\mathbf{T}$ is now called the *stress tensor*.

Using the equation

$$\mathbf{t}_{\mathbf{e}_1} = \mathbf{T}(\mathbf{e}_1) = T_{11}\mathbf{e}_1 + T_{21}\mathbf{e}_2 + T_{31}\mathbf{e}_3 \qquad (2.46)$$

we see that T_{11} is the normal component of the stress vector $\mathbf{t}_{\mathbf{e}_1}$, called the normal stress, while T_{21} and T_{31} are the shearing components of $\mathbf{t}_{\mathbf{e}_1}$, called the shear stresses. Similarly, we can interpret the components of $\mathbf{t}_{\mathbf{e}_2}$ and $\mathbf{t}_{\mathbf{e}_3}$. Therefore, the diagonal elements T_{11}, T_{22}, and T_{33} are called the *normal stresses*, while the off-diagonal elements T_{21}, T_{31}, T_{12}, T_{32}, T_{23}, and T_{13} are called the *shear stresses*.

Using equation (2.39), we can deduce the stress tensor components T_{ij} as follows:

$$t_i = T_{ij}n_j \qquad (2.47)$$

In matrix form, we can write equation (2.47) as follows:

$$[\mathbf{t}] = [\mathbf{T}][\mathbf{n}] \qquad (2.48)$$

Therefore, the state of stress at point P is completely characterized by the stress tensor **T**. In addition, the stress tensor **T** can be represented by a matrix as follows:

$$\mathbf{T} \equiv \begin{bmatrix} T_{11} & T_{12} & T_{13} \\ T_{21} & T_{22} & T_{23} \\ T_{31} & T_{32} & T_{33} \end{bmatrix} \tag{2.49}$$

The stress tensor can be shown to be symmetric using the principle of moment of momentum (see Problem 2.37). Therefore, in this text we will always use $T_{12} = T_{21}$, $T_{13} = T_{31}$, and $T_{23} = T_{32}$.

We know from Chapter 1 that for any symmetric tensor, there exists at least three mutually perpendicular principal directions corresponding to the eigenvalues of the tensor. In case of the stress tensor, the planes having these directions as their normals are called the *principal planes*. The normal stresses on these planes are called the *principal stresses*. Thus, the eigenvalues of **T** are the principal stresses which include the maximum and minimum values of normal stress among all planes passing through a given point.

Next, we derive the differential equations of motion. Consider a small rectangular element that is isolated from the continuum in the neighborhood of the position x_i, as shown in Figure 2.6. The vectors shown in the figure are the stress vectors on each face of the cube. Let $\mathbf{B} = B_i \mathbf{e}_i$ be the body force per unit mass, ρ be the mass density at x_i, and $\mathbf{a}$ the acceleration of a particle at x_i.

Using Newton's law of motion, we obtain:

$$\begin{aligned} [& \frac{\mathbf{t}_{\mathbf{e}_1}(x_1 + \Delta x_1, x_2, x_3) - \mathbf{t}_{\mathbf{e}_1}(x_1, x_2, x_3)}{\Delta x_1} \\ &+ \frac{\mathbf{t}_{\mathbf{e}_2}(x_1, x_2 + \Delta x_2, x_3) - \mathbf{t}_{\mathbf{e}_2}(x_1, x_2, x_3)}{\Delta x_2} \\ &+ \frac{\mathbf{t}_{\mathbf{e}_3}(x_1, x_2, x_3 + \Delta x_3) - \mathbf{t}_{\mathbf{e}_3}(x_1, x_2, x_3)}{\Delta x_3}] \Delta x_1 \Delta x_2 \Delta x_3 \\ &+ \rho \mathbf{B} \Delta x_1 \Delta x_2 \Delta x_3 = \rho \mathbf{a} \Delta x_1 \Delta x_2 \Delta x_3 \end{aligned} \tag{2.50}$$

Dividing equation (2.50) by $\Delta x_1 \Delta x_2 \Delta x_3$ and letting $\Delta x_i \longrightarrow 0$, we obtain:

$$\frac{\partial \mathbf{t}_{\mathbf{e}_1}}{\partial x_1} + \frac{\partial \mathbf{t}_{\mathbf{e}_2}}{\partial x_2} + \frac{\partial \mathbf{t}_{\mathbf{e}_3}}{\partial x_3} + \rho \mathbf{B} = \rho \mathbf{a} \tag{2.51}$$

Substituting $\mathbf{t}_{\mathbf{e}_j} = \mathbf{T}(\mathbf{e}_j) = T_{ij} \mathbf{e}_i$, we have:

$$\frac{\partial T_{ij}}{\partial x_j} \mathbf{e}_i + \rho B_i \mathbf{e}_i = \rho a_i \mathbf{e}_i \tag{2.52}$$

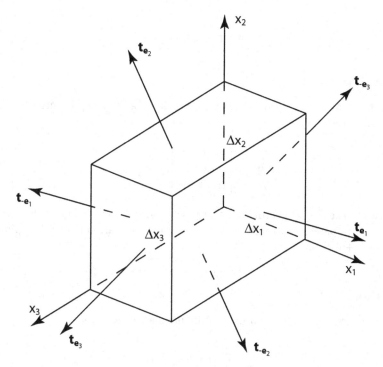

FIGURE 2.6
Infinitesimal cube with stress vectors

Equation (2.52) can be written as follows after dropping the indicial notation:

$$div\mathbf{T} + \rho\mathbf{B} = \rho\mathbf{a} \qquad (2.53)$$

Eliminating $\mathbf{e}_i$ from equation (2.52), we can rewrite it in component form as follows:

$$\frac{\partial T_{ij}}{\partial x_j} + \rho B_i = \rho a_i \qquad (2.54)$$

Equation (2.54) is called *Cauchy's equation of motion*. It must be satisfied for any continuum. If the acceleration vanishes, i.e., $\mathbf{a} = \mathbf{0}$, then we obtain the equilibrium equation as follows:

$$\frac{\partial T_{ij}}{\partial x_j} + \rho B_i = 0 \qquad (2.55)$$

or equivalently,

$$div\mathbf{T} + \rho\mathbf{B} = 0 \qquad\qquad (2.56)$$

Example 2.5

Given the state of stress at a certain point in a body is represented by the following stress tensor:

$$\mathbf{T} \equiv \begin{bmatrix} 1 & 2 & 0 \\ 2 & 3 & -1 \\ 0 & -1 & 4 \end{bmatrix} \text{MPa}$$

Determine the stress vector and the magnitude of the normal stress on a plane that passes through the point and is parallel to the plane $x + y - z + 4 = 0$.

Solution

The plane $x + y - z + 4 = 0$ has a unit normal $\mathbf{n}$ given by:

$$\mathbf{n} = \frac{1}{\sqrt{3}}(\mathbf{e}_1 + \mathbf{e}_2 - \mathbf{e}_3)$$

Therefore, we obtain the stress vector using equation (2.48) as follows:

$$[\mathbf{t}] = [\mathbf{T}][\mathbf{n}] = \frac{1}{\sqrt{3}}\begin{bmatrix} 1 & 2 & 0 \\ 2 & 3 & -1 \\ 0 & -1 & 4 \end{bmatrix}\begin{bmatrix} 1 \\ 1 \\ -1 \end{bmatrix} = \frac{1}{\sqrt{3}}\begin{bmatrix} 3 \\ 6 \\ -5 \end{bmatrix}$$

$$\mathbf{t} = \frac{1}{\sqrt{3}}(3\mathbf{e}_1 + 6\mathbf{e}_2 - 5\mathbf{e}_3)$$

The normal stress is now obtained as follows:

$$T_n = \mathbf{t} \cdot \mathbf{n} = \frac{1}{3}(3 + 6 + 5) = \frac{14}{3} = 4.67 \,\text{MPa}.$$

Example 2.6

A case of *plane stress* is defined as a state of stress such that $T_{13} = T_{23} = T_{33} = 0$. For this case of plane stress, determine the principal stresses and the corresponding principal directions.

Solution

In this case, the stress tensor is written as follows:

$$\mathbf{T} \equiv \begin{bmatrix} T_{11} & T_{12} & 0 \\ T_{12} & T_{22} & 0 \\ 0 & 0 & 0 \end{bmatrix}$$

where we have used $T_{21} = T_{12}$ because $\mathbf{T}$ is symmetric. The three stress invariants I_1, I_2, and I_3 are obtained as follows:

$$I_1 = T_{11} + T_{22}$$

$$I_2 = \begin{vmatrix} T_{11} & T_{12} \\ T_{12} & T_{22} \end{vmatrix} = T_{11}T_{22} - T_{12}^2$$

$$I_3 = |\mathbf{T}| = 0$$

Therefore, the characteristic equation of $\mathbf{T}$ is as follows:

$$\lambda^3 - I_1\lambda^2 + I_2\lambda - I_3 = 0$$

$$\lambda^3 - (T_{11} + T_{22})\lambda^2 + (T_{11}T_{22} - T_{12}^2)\lambda = 0$$

$$\lambda[\lambda^2 - (T_{11} + T_{22})\lambda + (T_{11}T_{22} - T_{12}^2)] = 0$$

The three eigenvalues are now given as the solutions λ for the above equation:

$$\lambda_1 = 0$$

$$\lambda_2 = \frac{T_{11} + T_{22} + \sqrt{(T_{11} + T_{22})^2 - 4(T_{11}T_{22} - T_{12}^2)}}{2}$$

$$\lambda_3 = \frac{T_{11} + T_{22} - \sqrt{(T_{11} + T_{22})^2 - 4(T_{11}T_{22} - T_{12}^2)}}{2}$$

In order to find the corresponding eigenvectors, we set $(T_{ij} - \lambda\delta_{ij})n_j = 0$ and obtain n_j for each one of the three eigenvalues. The three eigenvectors are then obtained as follows:

$$\mathbf{n_1} = \mathbf{e_3}$$

$$n_{2,3} = \cos\theta\mathbf{e_1} + \sin\theta\mathbf{e_2}$$

where

$$\tan\theta = -\frac{T_{11} - \lambda}{T_{12}}$$

Example 2.7

Consider the following stress distribution:

$$T_{11} = x_1^2 + x_2^2$$

$$T_{12} = -x_1 x_2$$

$$T_{22} = x_1^2 - x_2^2$$

$$T_{13} = 0$$

$$T_{23} = 0$$

$$T_{33} = x_1^2 x_2^2$$

Does the above stress distribution satisfy the equations of equilibrium. Assume there are no body forces.

Solution

Using equation (2.55) with $i = 1$, we obtain:

$$\frac{\partial T_{1j}}{\partial x_j} + \rho B_1 = 0$$

but $B_1 = 0$:

$$\frac{\partial T_{11}}{\partial x_1} + \frac{\partial T_{12}}{\partial x_2} + \frac{\partial T_{13}}{\partial x_3} = 0$$

$$2x_1 + (-x_1) + 0 = x_1 \neq 0$$

Therefore, the stress distribution given does not satisfy the equations of equilibrium.

Problems

2.33. Consider the following state of stress given by the stress tensor:

$$\mathbf{T} \equiv \begin{bmatrix} 1 & 2 & 1 \\ 2 & 3 & 5 \\ 1 & 5 & 2 \end{bmatrix} \text{ MPa}$$

Determine on which plane of the three coordinate planes (with normals e_1, e_2, e_3) the normal stress is greatest.

2.34. Consider the following state of stress given by the stress tensor:

$$\mathbf{T} \equiv \begin{bmatrix} 3 & 4 & -1 \\ 4 & 5 & 2 \\ -1 & 2 & 5 \end{bmatrix} \text{MPa}$$

  (a) Determine the stress vector at the point on the plane whose normal is in the direction of the vector $e_1 + e_2 + e_3$.

  (b) Determine the magnitude of the normal and shear stress on this plane.

2.35. Solve problem 2.34 again for a plane which is parallel to the plane $x_1 - 2x_2 + 3x_3 = 4$.

2.36. For any stress tensor $\mathbf{T}$, define the *deviatoric* stress tensor $\mathbf{T}_D$ by the equation:

$$\mathbf{T}_D = \mathbf{T} - \frac{1}{3}T_{kk}\mathbf{I}$$

  (a) Show that $(\mathbf{T}_D)_{mm} = 0$

  (b) Determine the deviatoric stress tensor T_D for the following stress tensor:

$$\mathbf{T} \equiv 20 \begin{bmatrix} 1 & 1 & 1 \\ 1 & 2 & 3 \\ 1 & 3 & 5 \end{bmatrix} \text{MPa}$$

2.37. Show that the stress tensor is symmetric. *Hint*: Use the principle of moment of momentum on a differential parallelpiped isolated from the body.

2.38. Consider the state of *hydrostatic pressure*. This state is defined as the state of stress at a certain point such that there is no shear stress on any plane passing through the point. Write a general equation for the stress tensor $\mathbf{T}$ for this state of hydrostatic pressure.

2.39. Consider the state of stress given in Example 2.5. Determine the value of T'_{12} given that $e'_1 = \frac{1}{2}(e_1 + e_2 + e_3)$ and $e'_2 = \frac{1}{2}(e_1 - e_2 - e_3)$

2.40. Show that the maximum shear stress is equal to one-half the difference between the maximum and the minimum principal stresses and the acts on the plane that bisects the angle between the directions of the maximum and minimum principal stresses.

2.41. Consider the state of plane stress given in Example 2.6. Determine the maximum shear stress.

2.42. Solve Example 2.6 again for the following state of stress:

$$\mathbf{T} \equiv \begin{bmatrix} 10 & 5 & 0 \\ 5 & 15 & 0 \\ 0 & 0 & 0 \end{bmatrix} \text{ MPa}$$

2.43. Use the state of stress given in Problem 2.42 to solve Problem 2.41 again.

2.44. Give an example of a stress tensor that satisfies the equations of equilibrium.

2.45. Does the following state of stress satisfy the equations of equilibrium. Assume there are no body forces.

$$\mathbf{T} \equiv \begin{bmatrix} x_1 + x_2 & -x_1 x_2 & 0 \\ -x_1 x_2 & x_1 - x_2 & 0 \\ 0 & 0 & x_1 x_2 \end{bmatrix}$$

2.46. Consider the following state of stress:

$$T_{11} = x_1^2 + \alpha x_2^2$$

$$T_{12} = -\alpha x_1 x_2$$

$$T_{13} = 0$$

$$T_{22} = x_1^2 - \beta x_2^2$$

$$T_{23} = 0$$

$$T_{33} = \gamma x_1 x_2$$

Determine the values of α, β, and γ such that the stress tensor $\mathbf{T}$ satisfies the equations of equilibrium. Assume that the body forces vanish.

2.47. Define an octahedral stress plane to be a plane which makes equal angles with each of the principal axes of stress.

(a) Show that the normal stress on an octahedral plane is given by one-third the first stress invariant.

(b) Show that the shear stress on the octahedral plane is given by:

$$T_s = \frac{1}{3}[(T_1 - T_2)^2 + (T_2 - T_3)^2 + (T_1 - T_3)^2]^{\frac{1}{2}}$$

where T_1, T_2, and T_3 are the principal stresses.

2.48. Can the following two matrices represent the same stress tensor in two different coordinate systems:

$$T_1 \equiv \begin{bmatrix} 10 & 20 & 0 \\ 20 & 10 & -5 \\ 0 & -5 & 30 \end{bmatrix} \text{ MPa}$$

$$T_2 \equiv \begin{bmatrix} 10 & 10 & 10 \\ 10 & 30 & 0 \\ 10 & 0 & 30 \end{bmatrix} \text{ MPa}$$

2.49. Consider the state of stress called *simple shear* where the only non-vanishing stress components are a single pair of shear stresses. For example $T_{12} = T_{21} = \tau$ and all other $T_{ij} = 0$.

(a) Determine the principal stresses and principal directions for this case of stress.

(b) Determine the maximum shear stress and the plane on which it acts for this case of stress.

2.50. Consider the state of stress with only the normal components nonvanishing such that $T_{11} = \sigma_1$, $T_{22} = \sigma_2$, and $T_{33} = \sigma_3$ with $\sigma_1 > \sigma_2 > \sigma_3$ and all other $T_{ij} = 0$. Determine the maximum shear stress and the plane on which it acts.

2.51. Show that the stress tensor is generally not symmetric if there are body moments per unit volume, as in the case of a polarized anisotropic dielectric solid.

2.52. Consider the stress tensor represented by the following matrix:

$$T \equiv \begin{bmatrix} \alpha & 10 & 0 \\ 10 & \beta & 30 \\ 0 & 30 & \gamma \end{bmatrix} \text{ MPa}$$

If the principal values for this stress tensor are $T_1 = 40$ MPa, $T_2 = 25$ MPa, and $T_3 = 5$ MPa, determine the values of α, β, and γ.

2.53. Consider the case of plane stress where the stress tensor is generally given as follows:

$$T_{11} = T_{11}(x_1, x_2)$$

$$T_{12} = T_{12}(x_1, x_2)$$

$$T_{13} = 0$$

$$T_{21} = T_{21}(x_1, x_2)$$

$$T_{22} = T_{22}(x_1, x_2)$$

$$T_{23} = 0$$

$$T_{31} = 0$$

$$T_{32} = 0$$

$$T_{33} = 0$$

Write the equation of equilibrium in this special case.

2.54. Consider the special case of the stress given in Problem 2.53. Consider also a scalar function $\phi(x_1, x_2)$ such that:

$$T_{11} = \frac{\partial^2 \phi}{\partial x_2^2}$$

$$T_{22} = \frac{\partial^2 \phi}{\partial x_1^2}$$

$$T_{12} = \frac{\partial^2 \phi}{\partial x_1 \partial x_2}$$

Does this stress distribution satisfy the equations of equilibrium assuming zero body forces.

2.55. Derive the equations of equilibrium in cylindrical coordinates. *Hint:* Consider a differential volume of material bounded by the three pairs of faces $r = r_o$, $r = r_o + dr$, $\theta = \theta_o$, $\theta = \theta_o + d\theta$, and $z = z_o$, $z = z_o + dz$, as shown in Figure 2.7.

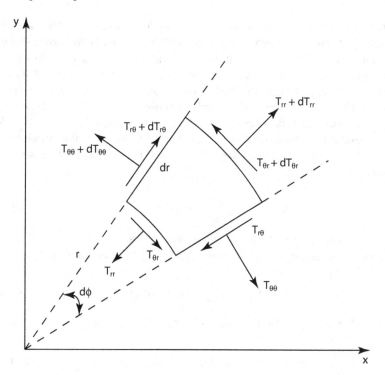

FIGURE 2.7
Infinitesimal element with stresses in cylindrical coordinates.(Lai et al., 1984)

2.4 Linear Elastic Relation

In this section we study the relation between stress and strain in the context of linear elasticity. A constitutive equation relates the stress to relevant quantities of deformation. We assume that deformations are small and the rate of load application has no effect. In general, we can, therefore, write:

$$\mathbf{T} = \mathbf{T}(\mathbf{E}) \tag{2.57}$$

with $\mathbf{T}(\mathbf{0}) = \mathbf{0}$. We assume the function $\mathbf{T}(\mathbf{E})$ to be linear in this section, so we can write the general linear constitutive relation as follows:

$$T_{ij} = C_{ijkl}E_{kl} \tag{2.58}$$

where C_{ijkl} are the components of a fourth-order tensor called the *elasticity tensor*. This tensor is characteristic of anisotropic Hookean elastic solids.

A material is called *isotropic* if its mechanical properties are independent of direction. On the other hand, a material is called *anisotropic* if its mechanical properties are dependent on direction. Therefore, for an isotropic material, the components of the elasticity tensor C_{ijkl} must be independent of the rotation and reflection of the rectangular basis, i.e., $C_{ijkl} = C'_{ijkl}$ for all orthogonal transformations of basis. Such a tensor is called an *isotropic tensor*. In general, C_{ijkl} can be written in terms of isotropic tensors as follows:

$$C_{ijkl} = \lambda \delta_{ij}\delta_{kl} + \alpha \delta_{ik}\delta_{jl} + \beta \delta_{il}\delta_{jk} \qquad (2.59)$$

where $\delta_{ij}\delta_{kl}$, $\delta_{ik}\delta_{jl}$, and $\delta_{il}\delta_{jk}$ are isotropic tensors.

Substituting equation (2.59) into equation (2.58), we obtain:

$$T_{ij} = \lambda e \delta_{ij} + (\alpha + \beta)E_{ij} \qquad (2.60)$$

$e = E_{kk}$. Let $\alpha + \beta = 2\mu$, then equation (2.60) becomes:

$$T_{ij} = \lambda e \delta_{ij} + 2\mu E_{ij} \qquad (2.61a)$$

or

$$\mathbf{T} = \lambda e \mathbf{I} + 2\mu \mathbf{E} \qquad (2.61b)$$

The two material constants λ and μ are called *Lame's constants*. In expanded form, equations (2.61) are the constitutive equations for linear isotropic elastic solids, given as follows:

$$T_{11} = \lambda(E_{11} + E_{22} + E_{33}) + 2\mu E_{11}$$

$$T_{22} = \lambda(E_{11} + E_{22} + E_{33}) + 2\mu E_{22}$$

$$T_{33} = \lambda(E_{11} + E_{22} + E_{33}) + 2\mu E_{33}$$

$$T_{12} = 2\mu E_{12}$$

$$T_{13} = 2\mu E_{13}$$

$$T_{23} = 2\mu E_{23}$$

Example 2.8

Consider the case of *uniaxial tension*. We define the ratio T_{11}/E_{11} as *Young's modulus* or *modulus of elasticity*, denoted by E_y. Show that the following relation holds between the modulus of elasticity E_y, Lame's constants and Poisson's ratio:

$$\mu = \frac{E_y}{2(1 + \nu)}$$

Solution

First, solve equation (2.61a) for the strain components E_{ij} in terms of the stress components T_{ij}. We obtain:

$$E_{ij} = \frac{1}{2\mu} \left[T_{ij} - \frac{\lambda}{3\lambda + 2\mu} T_{kk} \delta_{ij} \right] \qquad (2.62)$$

Alternatively, in expanded form we have:

$$E_{11} = \frac{1}{2\mu} \left[T_{11} - \frac{\lambda}{3\lambda + 2\mu} T_{11} \right] = \frac{\lambda + \mu}{\mu(3\lambda + 2\mu)} T_{11}$$

$$E_{22} = E_{33} = -\frac{\lambda}{2\mu(3\lambda + 2\mu)} T_{11} = -\frac{\lambda}{2(\lambda + \mu)} E_{11}$$

$$E_{12} = E_{13} = E_{23} = 0$$

where we have used $T_{22} = T_{33} = T_{13} = T_{23} = T_{12} = 0$ for the case of uniaxial tension.

Therefore, the ratio T_{11}/E_{11}, denoted by E_y, is obtained as follows:

$$E_y = \frac{T_{11}}{E_{11}} = \frac{\mu(3\lambda + 2\mu)}{\lambda + \mu} \qquad (2.63)$$

Also, we obtained the following expression for Poisson's ratio ν, defined as $-E_{22}/E_{11} = -E_{33}/E_{11}$.

$$\nu = -\frac{E_{22}}{E_{11}} = -\frac{E_{33}}{E_{11}} = \frac{\lambda}{2(\lambda + \mu)} \qquad (2.64)$$

Now, eliminate λ from equations (2.63) and (2.64) to obtain:

$$\lambda = \frac{E_y}{2(1 + \nu)} \qquad (2.65)$$

Example 2.9

Consider the case of *simple shear*, which corresponds to a state of stress where all the stress components vanish except one pair of off-diagonal elements, like T_{12} and T_{21} with $T_{12} = T_{21}$.

Define the *shear modulus* G as the ratio $T_{12}/(2E_{12})$. Show that the shear modulus G is equal to Lame's constant μ.

Solution

Using the equation (2.61a), we have:

$$T_{12} = 2\mu E_{12}$$

$$\therefore \ \mu = \frac{T_{12}}{2E_{12}} = G$$

It should be noted that the shear modulus G is defined as the ratio of the shearing stress in simple shear to the decrease in angle between elements that are initially in the $e_{1}-$ and $e_{2}-$ directions.

Problems

2.56. Given the following strain matrix:

$$\mathbf{E} \equiv \begin{bmatrix} 10 & 20 & 10 \\ 20 & 50 & 0 \\ 10 & 0 & 30 \end{bmatrix} \times 10^{-6}$$

Find the components of stress T_{ij} where Lame's constants are given as $\lambda = 119.2$ GPa and $\mu = 79.2$ GPa.

2.57. Show that the principal directions of stress and strain coincide for an isotropic Hookean material.

2.58. Determine the relation between the first invariants of stress and strain for an isotropic elastic material.

2.59. Consider the case of *hydrostatic stress* defined by $\mathbf{T} - \sigma \mathbf{I}$ where σ is the constant magnitude of stress. Define the *bulk modulus* K as the ratio of the hydrostatic normal stress σ, to the unit volume change $\Delta V/V = e$. Show that

$$K = \frac{2\mu + 3\lambda}{3}$$

2.60. (a) If for a specific material the ratio of the bulk modulus (see Problem 2.59) to Young's modulus is very large, find the approximate value of Poisson's ratio.

(b) Indicate why the material of part(a) above can be called *incompressible*.

2.61. Given Young's modulus $E_y = 70$ GPa and Poisson's ratio $\nu = 0.25$, determine Lame's constants λ and μ.

2.62. Given the data of Problem 2.61, determine the bulk modulus (see Problem 2.59).

2.63. Given the following state of stress at a point:

$$\mathbf{T} = \begin{bmatrix} 70 & 40 & 30 \\ 40 & 50 & 10 \\ 30 & 10 & 40 \end{bmatrix} \text{MPa}$$

Determine the strain components E_{ij} if $E_y = 100$ GPa, $G = \mu = 40$ GPa, and $\nu = 0.30$.

2.64. Consider an incompressible material where $\nu = 1/2$ (see Problem 2.60).

(a) Show that $\mu = E_y/3$, $\lambda = \infty$, $K = \infty$.

(b) Show that Hooke's law becomes:

$$\mathbf{T} = 2\mu\mathbf{E} + \frac{1}{3}T_{kk}\mathbf{I}$$

2.65. Show that the following relations hold among the elastic constants:

$$\lambda = \frac{E\nu}{(1+\nu)(1-2\nu)}$$

$$G = \frac{3KE}{9K - E}$$

$$\nu = \frac{3K - E}{6K}$$

$$E = \frac{9K(K - \lambda)}{3K - \lambda}$$

$$\frac{G}{\lambda + G} = 1 - 2\nu$$

$$\frac{\lambda}{\lambda + 2G} = \frac{\nu}{1 - \nu}$$

See Problem 2.59 and Example 2.9 for the definitions of K and G, respectively.

2.66. Show that when temperature effects are considered, equation (2.61a) is generalized as follows:

$$T_{ij} = \lambda E_{kk}\delta_{ij} + 2\lambda E_{ij} - \beta(T - T_o)\delta_{ij}$$

where T_o is the initial temperature when the body has no stress, T is the new temperature and β is a material constant.

2.67. Consider the case of compression of a rectangular parallelopiped defined by the following stress matrix:

$$\mathbf{T} \equiv \begin{bmatrix} -p & 0 & 0 \\ 0 & -p & 0 \\ 0 & 0 & -p \end{bmatrix}$$

where p > 0 is the magnitude of the inward normal stress. Find expressions for the dilatation and bulk modulus in this case.

2.68. Determine the relation between the principal values of the stress and strain tensors.

2.69. In general, show that if A_{ij} is a symmetric tensor and a is a scalar. then A_{ij} and $A_{ij} + a\delta_{ij}$ have the same principal axes. Determine also the relation between the principal values of the two tensors.

2.70. Show that the following relations hold among the elastic constants:

$$\lambda = \frac{3K\nu}{1 + \nu}$$

$$\mu = \frac{3KE}{9K - E}$$

$$E = 3K(1 - 2\nu)$$

3

Isotropic Damage Mechanics

3.1 Introduction

Damage in metals is mainly the process of the initiation and growth of micro-cracks and cavities. At that scale the phenomenon is discontinuous. Kachanov (1958) was the first to introduce a continuous variable related to the density of such defects. This variable has constitutive equations for evolution, written in terms of stress or strain, which may be used in structural calculations in order to predict the initiation of macro-cracks. The three main types of damage are:

1. *Ductile damage*: the constitutive equations for this type of damage have been formulated by Lemaitre (1984, 1986), Lemaitre and Dufailly (1987), and Voyiadjis and Kattan (1992). Both elastic and elasto-plastic damage effects were considered.

2. *Fatigue damage*: the constitutive equations for this type of damage have been formulated by Lemaitre (1971), Chaboche (1974), and Lemaitre and Chaboche (1974).

3. *Creep damage*: the constitutive equations for this type of damage were formulated by Leckie and Hayhurst (1974), Hult (1979), and Lemaitre and Chaboche (1974).

This chapter deals with isotropic damage mechanics where the assumption of isotropic damage is often sufficient to give a good prediction of the carrying capacity, the number of cycles or the time to local failure in structural components. In this case, the damage variable is scalar and the calculations are not too difficult because of the scalar nature of the damage variable. In Chapter 4 we will consider anisotropic damage mechanics where the damage variable is tensorial - see Chaboche (1981), Murakami and Ohno (1980), and Krajcinovic and Foneska (1981).

Structures, cracks, representative volume elements, macro-cracks, and micro-cracks and cavities represent different scales which have to be carefully defined. The following three definitions are outlined (Lemaitre, 1986).

1. The *micro-scale* is the scale at which the mechanisms of strain and damage may be described and understood. The following are three examples:

   (a) atoms for elasticity.

   (b) dislocations in crystals for plasticity in metals.

   (c) inclusions or micro-cracks for damage.

   It is at this scale that hypotheses are taken in order to write constitutive equations at the macro-scale.

2. The *macro-scale* is the scale of the representative volume element which is a mathematical representation (or point) small enough to define space partial derivative but large enough to consider that the discrete elementary mechanism of strain and damage are well represented by a mean leading to continuous variables. This scale is usually taken in the order of 0.1 mm for metals, 1 mm for polymers, 10 mm for wood, and 100 mm for concrete (Lemaitre, 1986).

   A macro-crack just initiated is a crack of that size and the phenomenological constitutive equations govern the behavior at that scale.

3. The *structure scale* is the scale of mechanical components (mm, cm, dm, m) for which a crack is of the order of one to several millimeters or centimeters.

It should be noted that stress, strain or damage as results of structure calculations can describe phenomena only at the macro-scale and structure scale. Nevertheless, it may be an improvement in comparison with classical fracture mechanics which uses more global concepts. Strain energy release rate, contour integrals, and even stress intensity factors result from an overall energetic analysis of the cracked structure. These concepts have obtained enormous success in the prediction of crack behavior when the structures are two-dimensional (plane stress or plane strain), elastic, or in the range of small scale yielding or when the multi-dimensional loading is a proportional loading, and periodic if fatigue crack growth is involved.

For more sophisticated problems it is difficult to use classical fracture mechanics concepts mainly because of the following effects (Lemaitre, 1986):

1. Effects of large-scale yielding plasticity, overload retardation effects in fatigue, and ductile rupture due to large deformation.

2. Effects of time-dependent behavior, creep crack growth, and creep-fatigue interaction.

3. Three-dimensional effects, cracks loaded in mixed modes, and evolution of crack front shape when loading is non-proportional.

4. Effects of damage, short cracks in metals, multiple cracking in concrete, and delamination of composites.

In damage mechanics and using continuous concepts, the crack tip is a process zone in which damage increases until the rigidity and strength vanish. This gives rise to a continuous definition of a crack (at the structure scale). A *crack* is therefore defined as a flat zone of high gradients of rigidity and strength in which the critical conditions of damage have been reached. It is then considered that the evolution of the crack is the evolution of the damaged zone as calculated element by element with re-calculation of the state of stress and strain (Lemaitre, 1986). For more details, the reader is referred to Chaboche (1988 a,b), Chaboche and Lesne (1988), Chaboche (1993), Chow and Wang (1987 a,b, 1989).

3.2 Damage Variables

A damage variable can be defined within the context of the two categories of damage characteristics which are commonly used. The first category does not characterize the damage itself where the value of the damage variable determines only the damage. The second category introduces damage variables that are associated with a physical definition of damage. The first three items below correspond to the first category while the remaining three items correspond to the second category. In general, damage variables can be defined in terms of each of the following quantities (Lemaitre, 1986):

1. The stress: using a damage equivalent stress σ^* deduced from a thermodynamic approach of damage, where σ^* is given by (Lemaitre and Baptiste, 1982):

$$\sigma^* = \sigma_{eq.}\sqrt{\frac{2}{3}(1+v) + 3(1-2v)(\frac{\sigma_H}{\sigma_{eq.}})^2} \qquad (3.1)$$

where v is Poisson's ratio, $\sigma_{eq.}$ is the von Mises equivalent stress given by:

$$\sigma_{eq.} = \sqrt{\frac{3}{2}(\sigma_{ij} - \sigma_H\delta_{ij})(\sigma_{ij} - \sigma_H\delta_{ij})} \qquad (3.2)$$

and σ_H is the hydrostatic stress given by:

$$\sigma_H = \frac{1}{3}\sigma_{KK} \qquad (3.3)$$

The criterion for crack initiation is taken as $\sigma^* = \sigma_c$ where σ_c is a critical stress level dependent on the characteristics of each material.

2. The strain: using the accumulated plastic strain in the sense of von Mises as follows:

$$p = \int_0^t \sqrt{\frac{2}{3}\dot{\epsilon}_{ij}^p \dot{\epsilon}_{ij}^p}\, dt \qquad (3.4)$$

3. The plastic strain energy: used as follows:

$$w^p = \int_0^t \sigma_{ij}\dot{\epsilon}_{ij}^p\, dt \qquad (3.5)$$

4. The porosity: used as the relative volume of cavities for ductile damage as follows (Rousselier, 1981):

$$P = \frac{\delta v}{\delta V} \qquad (3.6)$$

where v is the volume of cavities and V is the total volume.

5. The radius of the cavities as demonstrated by Mudry (1983).

6. The relative area of micro-cracks and intersections of cavities in any plane oriented by its normal **n** as follows (Lemaitre and Chaboche, 1985):

$$\phi_{(\mathbf{n})} = \frac{\delta S_\phi}{\delta S} \qquad (3.7)$$

where S_ϕ is the area of micro-cracks and intersections and S is the total area of the cross-section. With a correction for micro-stress concentrations and interactions of defects this concept gives rise to a continuous variable suitable for continuum mechanics. For this reason, this is the definition of the scalar damage variable used in this book.

If damage is considered as isotropic (Kachanov, 1958), then the damage variable ϕ is adopted as a scalar. Otherwise, if anisotropic damage is considered, then a damage vector $\boldsymbol{\phi}$ (Krajcinovic and Foneska, 1981) or a second-order or a fourth-order damage tensor is employed (Betten, 1981; Chaboche, 1984; Leckie and Onat, 1980; Murakami and Ohno, 1981).

Restricting ourselves to isotropic damage in this chapter, then $\phi = 0$ characterizes the virgin (undamaged) state, while $\phi = \phi_c \leq 1$ characterizes the initiation of a macro-crack. The parameter ϕ_c is a critical value for the damage variable usually taken between 0.2 and 0.8 for engineering materials. See the reference by Lemaitre and Chaboche (1994).

3.3 Effective Stress

We will first state the *hypothesis of isotropy*. Isotropic damage consists of cracks and cavities with an orientation distributed uniformly in all directions. In this case, the damage variable does not depend on the orientation **n** and the damaged state is completely characterized by the scalar ϕ. In this chapter we will limit ourselves to the case of isotropic damage where $\phi_{\mathbf{n}} = \phi$ for each vector **n** (see Figure 3.1).

The introduction of a damage variable which represents a surface density of discontinuities in the material leads directly to the concept of effective stress, i.e., to the stress calculated over the section which effectively resists the forces.

Consider a damaged solid in which an element of finite volume has been isolated, of a sufficiently large size with respect to the inhomogeneities of the medium, and imagine that this element has been grossly enlarged (see Figure 3.1).

Let S be the area of a section of the volume element identified by its normal **n**. On this section, cracks and cavities which constitute the damage leave traces of different forms.

Let $\bar{S}$ be the effective area of resistance ($\bar{S} < S$) taking account of the area of these traces, stress concentrations in the neighborhood of geometric discontinuities, and the interactions between the neighboring defects. Let S_ϕ be the difference:

$$S_\phi = S - \bar{S} \tag{3.8}$$

It should be noted that S_ϕ is the total area of the defect traces corrected for stress concentrations and interactions. We define the scalar damage variable ϕ ($\phi \equiv \phi_{\mathbf{n}}$) as follows:

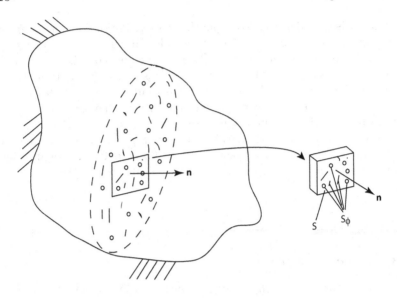

FIGURE 3.1

A damaged element showing the areas S and S_ϕ along with the normal vector **n**.

$$\phi = \frac{S_\phi}{S} \qquad\qquad (3.9)$$

From a physical point of view, the damage variable is, therefore, the relative (or corrected) area of cracks and cavities cut by a plane normal to the direction **n**. From the mathematical point of view, as S tends to 0, the variable ϕ (or $\phi_\mathbf{n}$) is the (corrected) surface density of the discontinuities of the matter in the plane normal to **n**.

In the uniaxial case, if F is the applied force on a section of the representative volume element, $\sigma = \frac{F}{S}$ is the usual stress satisfying equilibrium. In the presence of isotropic damage, ϕ, the effective area of resistance is obtained as follows using equations (3.8) and (3.9):

$$\bar{S} = S - S_\phi = S(1 - \phi) \qquad\qquad (3.10)$$

Taking $\sigma S = \bar{\sigma}\bar{S}$ along the two configurations, and substituting equation (3.10), we obtain the following equation for the effective stress $\bar{\sigma}$:

$$\bar{\sigma} = \frac{\sigma}{1 - \phi} \qquad (3.11)$$

It is clear that $\bar{\sigma} \geq \sigma$ and that $\bar{\sigma} = \sigma$ for a virgin material ($\phi = 0$), and $\bar{\sigma} \longrightarrow \infty$ at the moment of fracture ($\phi \longrightarrow 1$).

In the case of multi-axial isotropic damage, the ratio $S/\bar{S}$ does not depend on the orientation of the normal $\mathbf{n}$ and the operator $(1 - \phi)$ can be applied to all the components of the stress tensor. We will, therefore, have the following equation for the effective stress tensor $\bar{\sigma}$:

$$\bar{\sigma} = \frac{\sigma}{1 - \phi} \qquad (3.12)$$

We now state the *principle of strain equivalence*. We assume that the deformation behavior of the material is only affected by damage in the form of effective stress. Any deformation behavior, whether uniaxial or multi-axial, of a damaged material is represented by the constitutive laws of the virgin material in which the usual stress is replaced by the effective stress (see Figure 3.2).

For example, the uniaxial linear elastic law of a damaged material is written as follows:

$$\epsilon_e = \frac{\bar{\sigma}}{\bar{E}} = \frac{\sigma}{(1 - \phi)\bar{E}} \qquad (3.13)$$

where $\bar{E}$ is Young's modulus. This constitutes a nonrigorous hypothesis which assumes that all the different behaviors (elasticity, plasticity, viscoplasticity) are affected in the same way by the surface density of the damage defects. However, its simplicity allows the establishment of a coherent and efficient formalism.

By applying the concept of effective stress at the instant of fracture by interatomic decohesion, we define the critical value of damage ϕ_c, as that corresponding to the occurrence of this phenomenon.

If $\bar{\sigma}_u$ is the uniaxial stress at fracture by decohesion and σ_u is the usual ultimate fracture stress, we have (see equation (3.11)):

$$\bar{\sigma}_u = \frac{\sigma_u}{1 - \phi_c} \qquad (3.14)$$

Solving equation (3.14) for ϕ_c, we obtain:

$$\phi_c = 1 - \frac{\sigma_u}{\bar{\sigma}_u} \qquad (3.15)$$

The physics of solids shows that $\bar{\sigma}_u$ is of the order of $\bar{E}/50$ - $\bar{E}/20$; for common materials σ_u is of the order of $\bar{E}/100$ - $\bar{E}/250$, and ϕ_c is therefore of the order of 0.5 - 0.9. This allows us to neglect the term $(1 - \phi_c)^x$ with $x \gg 1$, a term which often appears in calculation, in comparison to 1.

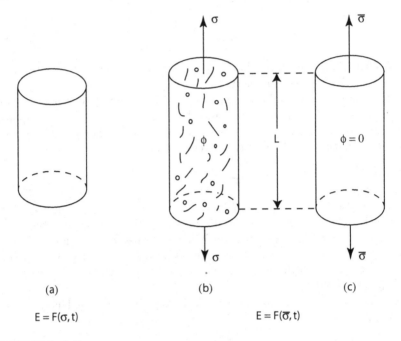

(a)   (b)   (c)

$E = F(\sigma, t)$ $E = F(\overline{\sigma}, t)$

FIGURE 3.2
Different configurations of the material: (a) virgin material, (b) damaged material, (c) equivalent virgin material. Note that the function F is the same used for all configurations.

Example 3.1

Let Y be the thermodynamic variable associated with the damage variable ϕ in the sense that $Y\dot{\phi}$ is the power dissipated in the damage process. Consider a thermodynamic potential Ψ quadratic with respect to the elastic strain ϵ_e and linear in $(1 - \phi)$. Show that:

$$Y = \frac{\partial \Psi}{\partial \phi} = -\frac{\sigma^{*2}}{E^2(1 - \phi)^2} \tag{3.16}$$

where σ^* is given in equation (3.1).

Solution

The thermodynamic potential Ψ is written as follows:

$$\Psi = (1 - \phi)\epsilon_e^2 \tag{3.17}$$

Substitute equation (3.13) into the above equation to obtain:

$$\Psi = \frac{\sigma^{*2}}{(1 - \phi)\bar{E}^2}$$

where σ is replaced by σ^* in the equation. Next, we take the derivative of Ψ with respect to ϕ to obtain:

$$Y = \frac{\partial \Psi}{\partial \phi} = -\frac{\sigma^{*2}}{(1 - \phi)^2 \bar{E}^2}$$

Example 3.2

Sidoroff (1981) proposed the *hypothesis of elastic energy equivalence* to be used instead of the hypothesis of strain equivalence. This new hypothesis assumes that the elastic energy for a damaged material is equivalent in form to that of the undamaged (effective) material except that the stress is replaced by the effective stress in the energy formulation. Use the hypothesis of elastic energy equivalence to show that:

$$\phi = 1 - \sqrt{\frac{E}{\bar{E}}} \tag{3.18}$$

where $\bar{E}$ is the effective modulus of elasticity and E is the modulus of elasticity of the damaged material.

Solution

For the uniaxial tension case, the constitutive relation is Hooke's law of linear elasticity given by:

$$\sigma = E\epsilon_e \tag{3.19}$$

The same linear elastic constitutive relation applies to the undamaged (effective) state, i.e.,

$$\bar{\sigma} = \bar{E}\bar{\epsilon}_e \tag{3.20}$$

where $\bar{\epsilon}_e$ and $\bar{E}$ are the effective counterparts of ϵ_e and E, respectively.

The hypothesis of elastic energy equivalence is written mathematically as follows:

$$\frac{1}{2}\sigma\epsilon_e = \frac{1}{2}\bar{\sigma}\bar{\epsilon}_e \tag{3.21}$$

where the elastic strain energy $\frac{1}{2}\sigma\epsilon_e$ is equated to the effective elastic strain energy $\frac{1}{2}\bar{\sigma}\bar{\epsilon}_e$.

Substituting equation (3.11) into equation (3.21) and simplifying, we obtain the following relation between the elastic strain ϵ_e and the effective elastic strain $\bar{\epsilon}_e$:

$$\bar{\epsilon}_e = (1 - \phi)\epsilon_e \qquad (3.22)$$

We next substitute equations (3.11) and (3.22) into equation (3.20), simplify the result and compare it with equation (3.19) to obtain:

$$E = \bar{E}(1 - \phi)^2 \qquad (3.23)$$

Equation (3.23) represents the transformation law for the modulus of elasticity. Solving equation (3.23) for ϕ, we finally obtain the desired result as follows:

$$\phi = 1 - \sqrt{\frac{E}{\bar{E}}}$$

Problems

3.1. Derive equation (3.13). Which hypothesis is used in the derivation.

3.2. Give an example for each one of the following damage types: (a) ductile damage, (b) fatigue damage, and (c) creep damage.

3.3. Why do you think the macro-scale is taken in the order of 0.1 mm for metals but 10 mm for wood.

3.4. Derive equation (3.1). *Hint*: See the paper by Lemaitre and Baptiste, 1982.

3.5. Define a new damage variable based on the radius of cavities. *Hint*: See the paper by Mudry, 1983.

3.6. Plot a graph showing the relation between $\bar{\sigma}$ and ϕ assuming that the stress σ is constant. See equation (3.11).

3.7. Why is the critical value of the damage variable ϕ_c usually taken between 0.5 and 0.9.

3.8. Compare the transformation laws for the elastic strain and the modulus of elasticity in the case of using (a) the hypothesis of strain equivalence, (b) the hypothesis of elastic energy equivalence.

3.9. How will equation (3.18) change if the hypothesis of elastic energy equivalence is replaced with the hypothesis of strain equivalence.

3.10. Suppose that the damage in a cylinder under uniaxial tension is composed entirely of voids and cracks. Let S^v be the area of voids and S^c be the area of cracks. The total area S is then given by:

$$S = \bar{S} + S^v + S^c$$

Show that, in this case, the damage variable ϕ is given by:

$$\phi = \phi^v + \phi^c - \phi^v \phi^c$$

where ϕ^v is the damage variable due to voids and ϕ^c is the damage variable due to cracks, both defined as follows:

$$\phi^v = \frac{S^v}{\bar{S} + S^v}$$

$$\phi^c = \frac{S^c}{\bar{S} + S^c}$$

3.11. Describe how equation (3.18) can be used to determine the damage variable ϕ experimentally.

3.12. Derive the transformation law for the compliance C where $C = 1/E$ in the case of uniaxial tension.

3.13. Show that the critical value of the damage variable ϕ_c can be calculated using the following formula:

$$\phi_c = 1 - \frac{\sigma^*}{\bar{E}\sqrt{-Y_c}}$$

where σ^* is given in equation (3.1), $\bar{E}$ is the modulus of elasticity, and Y_c is the critical value of the thermodynamic variable Y associated with the damage variable ϕ. See Example 3.1.

3.14. Consider the density of elastic strain energy W^e defined by:

$$dW^e = \sigma d\epsilon^e$$

where the operator "d" indicates an incremental quantity. Show that the thermodynamic variable Y associated with the damage variable ϕ is given by:

$$Y = -\frac{1}{2}\frac{dW^e}{d\phi}$$

3.4   Damage Evolution

There are several approaches in the literature on the topic of evolution of damage and the proper form of the kinetic equation of the damage variable. Kachanov (1986) proposed an evolution equation of damage based on a power law with two independent material constants. However, the resulting kinetic equation for the damage variable evolution is complicated and difficult to solve. Therefore, a more rational approach based on energy considerations will be adopted in this book.

The approach followed will depend on the introduction of a damage strengthening criterion in terms of a scalar function g, and a generalized thermodynamic force Y that corresponds to the damage variable ϕ (Lemaitre, 1985; Lee et al, 1985; Voyiadjis and Kattan, 1999). Substituting equations (3.20) and (3.22) into the right-hand-side of equation (3.21), we obtain the elastic strain energy U in the damaged state of the material as follows:

$$U = \frac{1}{2}\bar{E}(1 - \phi)^2\epsilon^{e2} \tag{3.24}$$

in which $\bar{E}$ is constant. Therefore, the incremental elastic strain energy dU is obtained by differentiating equation (3.24) as follows:

$$dU = \bar{E}(1 - \phi)^2\epsilon^e d\epsilon^e - \bar{E}(1 - \phi)\epsilon^{e2}d\phi \tag{3.25}$$

The generalized thermodynamic force Y associated with the damage variable ϕ is thus defined by:

$$Y = \frac{dU}{d\phi} = -\bar{E}(1 - \phi)\epsilon^{e2} \tag{3.26}$$

Let $g(Y, L)$ be the damage function (criterion) as proposed by Lee et al (1985), where $L \equiv L(l)$ is a damage strengthening parameter that is a function of the "overall" damage parameter l. In this case, the scalar function g takes the following form:

$$g(Y, L) = \frac{1}{2}Y^2 - L(l) \equiv 0 \tag{3.27}$$

The damage strengthening criterion defined by equation (3.27) is similar to the von Mises yield criterion in the theory of plasticity. In order to derive a normality rule for the evolution of damage, we first start with the power of dissipation Π which is given by:

$$\Pi = -Y d\phi - L dl \tag{3.28}$$

where the "d" in front of a variable indicates the incremental quantity of the variable. The problem is to extremize Π subject to the condition $g = 0$. Using the mathematical theory of functions of several variables, we introduce the Lagrange multiplier $d\lambda$, and form the objective function $\Psi(Y, L)$ such that:

$$\Psi = \Pi - d\lambda.g \qquad (3.29)$$

The problem reduces to extremizing the function Ψ. For this purpose, the two necessary conditions are $\partial\Psi/\partial Y = 0$ and $\partial\Psi/\partial L = 0$. Using these two conditions, along with equations (3.28) and (3.29), we obtain:

$$d\phi = -d\lambda \frac{\partial g}{\partial Y} \qquad (3.30a)$$

$$dl = -d\lambda \frac{\partial g}{\partial L} \qquad (3.30b)$$

Substituting for g from equation (3.27) into equation (3.30b), one concludes directly that $d\lambda = dl$. Substituting this into equation (3.30a), along with using equation (3.27), we obtain:

$$d\phi = -d\lambda.Y \qquad (3.31)$$

In order to solve the differential equation (3.31), we must first find an expression for the Lagrange multiplier $d\lambda$. This can be done by invoking the consistency condition $dg = 0$. Applying this condition to equation (3.27), we obtain:

$$\frac{\partial g}{\partial Y} dY + \frac{\partial g}{\partial L} dL = 0 \qquad (3.32)$$

Substituting for $\frac{\partial g}{\partial Y}$ and $\frac{\partial g}{\partial L}$ from equation (3.27) and for $dL = dl(\frac{\partial L}{\partial l})$ using the chain rule of differentiation, and solving for dl, we obtain:

$$dl = d\lambda = \frac{Y dY}{\partial L/\partial l} \qquad (3.33)$$

Substituting the above expression of $d\lambda$ into equation (3.31), we obtain the kinetic (evolution) equation of damage as follows:

$$d\phi = \frac{-1}{\partial L/\partial l} Y^2 dY \qquad (3.34)$$

with the initial condition that $\phi = 0$ when $Y = 0$.

Example 3.3

Derive the following relation between the damage variable ϕ and elastic strain ϵ^e:

$$\frac{\phi}{(1-\phi)^3} = \frac{\bar{E}^3}{3(\frac{\partial L}{\partial l})}\epsilon^{e^6} \tag{3.35}$$

Solution

Differentiate the expression of Y in equation (3.26) to obtain:

$$dY = \bar{E}\epsilon^e[\epsilon^e d\phi - 2d\epsilon^e(1-\phi)] \tag{3.36}$$

Next, substitute the expressions of Y and dY of equations (3.26) and (3.36), respectively, into equation (3.34), to obtain the strain-damage differential equation as follows:

$$d\phi = \frac{1}{(\partial L/\partial l)}\bar{E}^3\epsilon^{e^5}(1-\phi)^2[2d\epsilon^e(1-\phi) - \epsilon^e d\phi] \tag{3.37}$$

The above differential equation can be solved easily by the simple change of variables $x = \epsilon^{e^2}(1-\phi)$ and noting that the expression on the right-hand-side of the equation (3.37) is nothing but $\bar{E}^3 x^2 dx/(\partial L/\partial l)$. Perform the integration with the initial condition that $\phi = 0$ when $\epsilon^e = 0$ to obtain the desired formula:

$$\frac{\phi}{(1-\phi)^3} = \frac{\bar{E}^3}{3(\partial L/\partial l)}\epsilon^{e^6}$$

One should note that an initial condition involving an initial damage value ϕ° could have been used, i.e., $\phi = \phi^\circ$ when $\epsilon^e = 0$.

Problems

Determine an evolution equation of damage based on a power law with two independent material constants as proposed by Kachanov in 1986.

Solve the differential equation (3.34) for ϕ in terms of Y. Assume a linear function for $L(l)$ in the form $L(l) = cl + d$, where c and d are constants. Plot the resulting formula and show that you obtain a cubic curve.

Solve the differential equation (3.37) again but this time using the initial condition $\phi = \phi^\circ = constant$ when $\epsilon^e = 0$.

Use equations (3.26) and (3.34) to derive equation (3.35) directly. *Hint:* Use the result of Problem 3.16 above.

Plot the relation between the damage variable ϕ and elastic strain ϵ^e obtained in equation (3.35).

The equations of damage evolution given in Section 3.4 were obtained based on the hypothesis of elastic energy equivalence. Obtain alternative but similar equations of damage evolution using the hypothesis of strain equivalence.

Consider the following damage function (criterion) g which does not employ the thermodynamic force Y in the formulation:

$$g = \frac{1}{2}\bar{\sigma}^2 - L(l) \equiv 0 \tag{3.38}$$

Use this new function instead of equation (3.27) to derive a new damage evolution equation to be used instead of equation (3.34).

Obtain a new relation between the damage variable ϕ and elastic strain ϵ^e based on the new damage function (criterion) g given in Problem 3.21 above.

Consider a composite cylindrical bar made of a matrix and fibers of two different materials. Let E^M and E^F be the moduli of elasticity for the matrix and fibers, respectively. Let also σ^M and σ^F be the stresses in the matrix and fibers, respectively. Then the total uniaxial stress σ in the bar is given by the following rule of mixtures:

$$\sigma = c^M \sigma^M + c^F \sigma^F$$

where c^M and c^F are the matrix and fiber volume (or area) fractions, respectively.

3.1. Define two new scalar damage variables ϕ^M and ϕ^F, for the matrix and fibers, respectively.

3.2. Determine two appropriate evolution equations for the two damage variables defined in part (a).

3.3. Determine a relation between the global damage variable ϕ and the two local damage variables ϕ^M and ϕ^F.

4

Kinematic Description of Damage

4.1 Introduction

In 1958, Kachanov (1958) introduced the concept of effective stress in damaged materials. This pioneering work started the subject that is now known as continuum damage mechanics. Research in this area has steadily grown and reached a stage that warrants its use in today's engineering applications. Continuum damage mechanics is now widely used in different areas including brittle failure (Krajcinovic, 1983, 1985; Krajcinovic and Foneska, 1981; Lubarda et at., 1994; Ju and Lee, 1991; Lee and Ju, 1991; Ju and Chen, 1994a,b), ductile failure (Lemaitre, 1985, 1986; Chaboche, 1979, 1981, 1988a,b; Chow and Wang, 1987), composite materials (Allen et al., 1987; Boyd et al., 1983; Voyiadjis and Kattan, 1993, Voyiadjis and Park, 1995a, b) and fatigue (Chow and Wei, 1991). In this theory, a continuous damage variable is defined and used to represent degradation of the material which reflects various types of damage at the micro-scale level like nucleation and growth of voids, cavities, micro-cracks, and other microscopic defects.

In continuum damage mechanics, the effective stress tensor is usually not symmetric. This leads to a complicated theory of damage mechanics involving micropolar media and the Cosserat continuum. Therefore, to avoid such a theory, symmetrization of the effective stress tensor is used to formulate a continuum damage theory in the classical sense (Lee et al., 1986; Sidoroff, 1979; Cordebois and Sidoroff, 1982; Murakami and Ohno, 1980; Betten, 1983; Lu and Chow, 1990). A linear transformation tensor, defined as a fourth-order damage effect tensor is used to symmetrize the effective stress tensor.

The kinetics of damage is well defined presently through the effective stress concept. However, the kinematics of the deformation with damage is only considered indirectly and is only limited to the small strain theory based on the hypothesis of the strain equivalence (Lemaitre and Chaboche, 1978) or energy equivalence (Sideroff, 1980). The finite deformation damage models by Ju (1989) and Zbib (1993) emphasize that "added flexibility" due to the existence of microcracks or microvoids is already embedded in the deformation gradient implicitly. Murakami (1988) presented the kinematics of damage deformation using the second-order damage tensor. However, the lack of an explicit formulation for the kinematics of finite deformation with damage leads to the failure in obtaining an explicit derivation of the kinematics that directly consider the damage deformation.

The kinematics of damage is described here explicitly by considering damage directly in the kinematic field using the second-order damage tensor. The deformation gradient of damage is defined using the second-order damage tensor based on the geometically symmetrized effective stress concept (using a fourth-order tensor). The Green deformation tensor of the elastic damage deformation is also derived.

For a detailed review of the principles of continuum damage machanics as used in this chapter, the reader is referred to the works of Kachanov (1958), Lemaitre (1985, 1986), Krajcinovic (1985), Lubarda and Krajcinovic (1995), Chaboche (1981, 1988a, b), Murakami (1988), Sidoroff (1979, 1980), and Voyiadjis and Kattan (1992, 1999). The following sections are based mainly on the work of Park and Voyiadjis (1998).

4.2   Theoretical Preliminaries

Referring to Figure 4.1, the initial undeformed configuration of the body is denoted by C_0, while the elastic damage deformed configuration after the body is subjected to a set of external agencies is denoted by C. The body in configuration C_0 undergoes a sequence of deformations starting with an elastic deformation without damage, followed by a damage deformation. This is indicated by path I in Figure 4.1. The configuration denoted by C^e implies the elastic deformed configuration. The initial undeformed body may have a pre-existing damage state.

A fictitious effective configuration for the body denoted by $\bar{C}$ is assumed to be obtained from C by fictitiously removing all the damage that the body has undergone. This is the fictitious effective configuration which is based on the effective stress concept. In this configuration, the body has only deformed elastically without damage. In addition to the fictitious effective configuration $\bar{C}$, the initial fictitious effective configuration denoted by $\bar{C}_0$ is defined by removing the initial damage from the initial undeformed configuration of the body. In the case of no initial damage existing in the undeformed body, the initial fictitious effective configuration is identical to the initial undeformed configuration. This chapter is limited to elastic strains with damage.

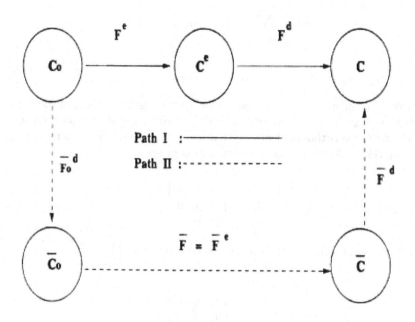

FIGURE 4.1
Schematic representation of elastic damage deformation configurations. (Park and Voyiadjis, 1998)

4.3   Description of Damage State

The damage state can be described using an even order tensor (Leckie, 1993; Onat, 1986; and Betten, 1986). Ju (1990) pointed out that even for isotropic damage one should employ a damage tensor (not a scalar damage variable) to characterize the state of damage in materials. However, the damage generally is anisotropic due to the external agency condition or the material nature itself. Although the fourth-order damage tensor can be used directly as a linear transformation tensor to define the effective stress tensor, it is not easy to characterize physically the fourth-order damage tensor compared to the second-order damage tensor. In this chapter, the damage is considered as a symmetric second-order tensor. The second-order damage tensor is given by (Murakami 1983) as follows:

$$\phi_{ij} = \sum_{k=1}^{3} \hat{\phi}_k \hat{n}_i^k \hat{n}_j^k \quad \text{(no sum in } k) \tag{4.1}$$

$$\phi_{ij} = b_{ir} b_{js} \hat{\phi}_{rs} \tag{4.2}$$

where $\hat{n}^k$ is an eigenvector corresponding to the eigenvalue, $\hat{\phi}_k$, of the damage tensor, ϕ. The damage tensor in the coordinate system that coincides with the three orthogonal principal directions of the damage tensor, $\hat{\phi}_{rs}$, in equation (4.2) is of diagonal form and is given by

$$\hat{\phi}_{ij} = \begin{bmatrix} \hat{\phi}_1 & 0 & 0 \\ 0 & \hat{\phi}_2 & 0 \\ 0 & 0 & \hat{\phi}_3 \end{bmatrix} \tag{4.3}$$

and the second order transformation tensor **b** is given by

$$b_{ir} = \begin{bmatrix} n_1^1 & n_2^1 & n_3^1 \\ n_1^2 & n_2^2 & n_3^2 \\ n_1^3 & n_2^3 & n_3^3 \end{bmatrix} \tag{4.4}$$

This proper orthogonal transformation tensor requires that

$$b_{ij} b_{kj} = \delta_{ik} \tag{4.5}$$

where δ_{ik} is a Kronecker delta and the determinant of the matrix $[\mathbf{b}]$ is given by

$$|[\mathbf{b}]| = 1 \tag{4.6}$$

Voyiadjis and Venson (1995) quantified the physical values of the eigenvalues $\hat{\phi}_k$ ($k = 1,2,3$) and the second-order damage tensor ϕ for the unidirectional fibrous composite by measuring the crack density with the assumption that one of the eigendirections of the damage tensor coincides with the fiber direction provided that the load is applied uniformly along the fiber direction. This introduces a distinct kinematic measure of damage which is complementary to the deformation kinematic measure of strain. A thermodynamically consistent evolution equation for the damage tensor ϕ together with a generalized thermodynamic force (Chaboche 1977) conjugate to the damage tensor is presented in the paper by Voyiadjis and Park (1995a,b).

4.4 Fourth-Order Anisotropic Damage Effect Tensor

In a general state of deformation and damage, the effective stress tensor $\bar{\sigma}$ is related to the stress tensor σ by the following linear transformation (Murakami and Ohno 1980):

$$\bar{\sigma}_{ij} = M_{ikjl}\sigma_{kl} \qquad (4.7)$$

$$\bar{\sigma} = \mathbf{M}\sigma \qquad (4.8)$$

where σ is the Cauchy stress tensor and $\mathbf{M}$ is a fourth-order linear transformation operator called the damage effect tensor. Depending on the form used for $\mathbf{M}$, it is very clear from equation (4.8) that the effective stress tensor $\bar{\sigma}$ is generally not symmetric. Using a nonsymmetric effective stress tensor as given by equation (4.8) to formulate a constitutive model will result in the introduction of the Cosserat and a micropolar continuum. However, the use of such complicated mechanics can be easily avoided if the proper fourth-order linear transformation tensor is formulated in order to symmetrize the effective stress tensor. Such a linear transformation tensor called the damage effect tensor is obtained in the literature (Lee at al., 1986; Sidoroff, 1979) using symmetrization methods. The effective stress tensor is symmetrized using the following expressions by Cordebois and Sidoroff (1979) and Lee et al. (1986).

$$\bar{\sigma}_{ij} = (\delta_{ik} - \phi_{ik})^{-1/2}\sigma_{kl}(\delta_{jl} - \phi_{jl})^{-1/2} \qquad (4.9)$$

$$\bar{\sigma}_{ij} = \tfrac{1}{2}[(\delta_{ik} - \phi_{ik})^{-1}\sigma_{kl}\delta_{jl} + \delta_{ik}\sigma_{kl}(\delta_{jl} - \phi_{jl})^{-1}] \qquad (4.10)$$

$$\sigma_{ij} = \tfrac{1}{2}[(\delta_{ik} - \phi_{ik})\bar{\sigma}_{kl}\delta_{jl} + \delta_{ik}\bar{\sigma}_{kl}(\delta_{jl} - \phi_{jl})] \qquad (4.11)$$

The fourth-order damage effect tensors corresponding to equations (4.9), (4.10) and (4.11) are defined such as

$$M_{ikjl} = (\delta_{ik} - \phi_{ik})^{-1/2}(\delta_{jl} - \phi_{jl})^{-1/2} \qquad (4.12)$$

$$M_{ikjl} = \tfrac{1}{2}[(\delta_{ik} - \phi_{ik})^{-1}\delta_{jl} + \delta_{ik}(\delta_{jl} - \phi_{jl})^{-1}] \qquad (4.13)$$

$$M_{ikjl} = 2[(\delta_{ik} - \phi_{ik})\delta_{jl} + \delta_{ik}(\delta_{jl} - \phi_{jl})]^{-1} \qquad (4.14)$$

Another approach is that of the damage effect tensor using the fourth order damage tensor ψ as defined by Chaboche (1979)

$$M_{ikjl} = (I_{ikjl} - \psi_{ikjl}) \tag{4.15}$$

where I_{ikjl} is a fourth order identity tensor and is given by

$$I_{ikjl} = \tfrac{1}{2}(\delta_{ij}\delta_{kl} + \delta_{il}\delta_{kj}) \tag{4.16}$$

However, it is not easy to characterize physically the fourth order damage tensor ψ_{ijkl} as opposed to the second-order damage tensor ϕ_{ij}. For the case of isotropic damage, the fourth order damage tensor is defined by Ju (1990) as follows

$$\psi_{ijkl} = d_1\delta_{ik}\delta_{jl} + d_2 I_{ikjl} \tag{4.17}$$

where d_1 and d_2 are scalars (dependent or independent) damage variables.

In order to describe the kinematics of damage, the physical meaning of the fourth-order damage effect tensor should be interpreted and not merely given as the symmetrization of the effective stress. In this chapter, the fourth-order damage effect tensor given by equation (4.12) will be used because of its geometrical symmetrization of the effective stress (Cordebois and Sidoroff, 1979). However, it is very difficult to obtain the explicit representation of $(\delta_{ik} - \phi_{ik})^{-1/2}$. The explicit representation of the fourth-order damage effect tensor $\mathbf{M}$ using the second-order damage tensor ϕ is of particular importance in the implementation of the constitutive modeling of damage mechanics. Therefore, the damage effect tensor $\mathbf{M}$ of equation (4.12) should be obtained using the coordinate transformation of the principal damage direction coordinate system. Thus the fourth-order damage effect tensor given by equation (4.12) can be written as

$$M_{ikjl} = b_{mi}b_{nj}b_{pk}b_{ql}\hat{M}_{mnpq} \tag{4.18}$$

where $\hat{\mathbf{M}}$ is a fourth-order damage effect tensor with reference to the principal damage direction coordinate system. The fourth-order damage effect tensor $\hat{\mathbf{M}}$ can be written as follows:

$$\hat{M}_{mpnq} = \hat{a}_{mp}\hat{a}_{nq} \tag{4.19}$$

where the second-order tensor a of the principal damage direction coordinate system is given by

$$\hat{a}_{mp} = [\delta_{mp} - \hat{\phi}_{mp}]^{-1/2}$$

$$= \begin{bmatrix} \dfrac{1}{\sqrt{1-\hat{\phi}_1}} & 0 & 0 \\[2mm] 0 & \dfrac{1}{\sqrt{1-\hat{\phi}_2}} & 0 \\[2mm] 0 & 0 & \dfrac{1}{\sqrt{1-\hat{\phi}_3}} \end{bmatrix} \tag{4.20}$$

Substituting equation (4.19) into equation (4.18), one obtains the following relation

$$M_{ikjl} = b_{mi}b_{nj}b_{pk}b_{ql}\hat{a}_{mp}\hat{a}_{nq}$$
$$= a_{ik}a_{jl} \tag{4.21}$$

Using equation (4.21), a second-order tensor **a** is defined as follows:

$$a_{ik} = b_{mi}b_{pk}\hat{a}_{mp} \tag{4.22}$$

The matrix form of equation (4.22) is as follows:

$$[a] = [b]^T[\hat{a}][b]$$

$$[a] = \begin{bmatrix} \dfrac{b_{11}b_{11}}{\sqrt{1-\hat{\phi}_1}} + \dfrac{b_{21}b_{21}}{\sqrt{1-\hat{\phi}_2}} + \dfrac{b_{31}b_{31}}{\sqrt{1-\hat{\phi}_3}} & \dfrac{b_{11}b_{12}}{\sqrt{1-\hat{\phi}_1}} + \dfrac{b_{21}b_{22}}{\sqrt{1-\hat{\phi}_2}} + \dfrac{b_{31}b_{32}}{\sqrt{1-\hat{\phi}_3}} \\[2ex] \dfrac{b_{12}b_{11}}{\sqrt{1-\hat{\phi}_1}} + \dfrac{b_{22}b_{21}}{\sqrt{1-\hat{\phi}_2}} + \dfrac{b_{32}b_{31}}{\sqrt{1-\hat{\phi}_3}} & \dfrac{b_{12}b_{12}}{\sqrt{1-\hat{\phi}_1}} + \dfrac{b_{22}b_{22}}{\sqrt{1-\hat{\phi}_2}} + \dfrac{b_{32}b_{32}}{\sqrt{1-\hat{\phi}_3}} \\[2ex] \dfrac{b_{13}b_{11}}{\sqrt{1-\hat{\phi}_1}} + \dfrac{b_{23}b_{21}}{\sqrt{1-\hat{\phi}_2}} + \dfrac{b_{33}b_{31}}{\sqrt{1-\hat{\phi}_3}} & \dfrac{b_{13}b_{12}}{\sqrt{1-\hat{\phi}_1}} + \dfrac{b_{23}b_{22}}{\sqrt{1-\hat{\phi}_2}} + \dfrac{b_{33}b_{32}}{\sqrt{1-\hat{\phi}_3}} \end{bmatrix}$$

$$\begin{bmatrix} \dfrac{b_{11}b_{13}}{\sqrt{1-\hat{\phi}_1}} + \dfrac{b_{21}b_{23}}{\sqrt{1-\hat{\phi}_2}} + \dfrac{b_{31}b_{33}}{\sqrt{1-\hat{\phi}_3}} \\[2ex] \dfrac{b_{12}b_{13}}{\sqrt{1-\hat{\phi}_1}} + \dfrac{b_{22}b_{23}}{\sqrt{1-\hat{\phi}_2}} + \dfrac{b_{32}b_{33}}{\sqrt{1-\hat{\phi}_3}} \\[2ex] \dfrac{b_{13}b_{13}}{\sqrt{1-\hat{\phi}_1}} + \dfrac{b_{23}b_{23}}{\sqrt{1-\hat{\phi}_2}} + \dfrac{b_{33}b_{33}}{\sqrt{1-\hat{\phi}_3}} \end{bmatrix} \tag{4.23}$$

4.5 Kinematic Description of Elastic-Damage Deformation

A schematic drawing representing the kinematics of finite elastic deformation and damage is shown in Figure 4.1. C_0 is the initial undeformed configuration of the body which may have initial damage in the material. C represents the final elastically deformed and damaged configuration of the body. The configuration $\bar{C}_0$ represents the initial configuration of the body that is obtained by fictitiously removing the initial damage from the C_0 configuration. If the initial configuration is undamaged then there is no difference between configurations C_0 and $\bar{C}_0$. Configuration $\bar{C}$ is obtained by fictitiously removing the damage from configuration C. Considering path I the deformation gradient referred to the undeformed configuration, C_0, denoted by **F** is polarly decomposed into the elastic deformation gradient denoted by $\mathbf{F}^e$ and the damage deformation gradient denoted by $\mathbf{F}^d$ such that

$$\mathbf{F} = \mathbf{F}^e\mathbf{F}^d \tag{4.24}$$

The Green deformation tensor of the elastic damage deformation can be obtained through either path I or path II as shown in Figure 4.1. Path I gives the Green deformation tensor as follows:

$$\begin{aligned} \mathbf{G} &= \mathbf{F}^T \mathbf{F} \\ &= \mathbf{F}^{d^T} \mathbf{F}^{e^T} \mathbf{F}^e \mathbf{F}^d \end{aligned} \tag{4.25}$$

Considering path II the fictitious initial damage Green deformation tensor, $\bar{\mathbf{G}}_0^d$, and the fictitious damage Green deformation tensor, $\bar{\mathbf{G}}^d$, are defined as follows:

$$\bar{\mathbf{G}}_0^d = \bar{\mathbf{F}}_0^{d^T} \bar{\mathbf{F}}_0^d \tag{4.26}$$

and

$$\bar{\mathbf{G}}^d = \bar{\mathbf{F}}^{d^T} \bar{\mathbf{F}}^d \tag{4.27}$$

respectively. $\bar{\mathbf{F}}_0^d$ and $\bar{\mathbf{F}}^d$ are the fictitious effective initial damage deformation and the fictitious effective damage deformation gradients, respectively. The deformation gradient and the Green deformation tensor used in following path II may appear initially to be of the following forms:

$$\mathbf{F} = \bar{\mathbf{F}}_0^d \bar{\mathbf{F}}^e \bar{\mathbf{F}}^d \tag{4.28}$$

and

$$\mathbf{G} = \bar{\mathbf{F}}^{d^T} \bar{\mathbf{F}}_0^{e^T} \bar{\mathbf{F}}_0^{d^T} \bar{\mathbf{F}}_0^d \bar{\mathbf{F}}^e \bar{\mathbf{F}}^d \tag{4.29}$$

respectively. However, due to the fictitious removal of damage from configurations C_0 and C, the deformation gradient and the Green deformation tensor cannot be expressed by equations (4.28) and (4.29). Since the two effective configurations, $\bar{C}_0$ and $\bar{C}$, are obtained by fictitiously removing damage from the real configurations, therefore the difference between the two fictitious damage deformation tensors needs to be subtracted from the Green deformation tensor given by equation (4.29). Thus, the resulting Green deformation tensor in considering path II is given by

$$\begin{aligned} \mathbf{G} &= \bar{\mathbf{F}}^{d^T} \bar{\mathbf{F}}_0^{e^T} \bar{\mathbf{F}}_0^{d^T} \bar{\mathbf{F}}_0^d \bar{\mathbf{F}}^e \bar{\mathbf{F}}^d - (\bar{\mathbf{G}}^d - \bar{\mathbf{G}}_0^d) \\ &= \bar{\mathbf{F}}^{d^T} \bar{\mathbf{F}}_0^{e^T} \bar{\mathbf{F}}_0^{d^T} \bar{\mathbf{F}}_0^d \bar{\mathbf{F}}^e \bar{\mathbf{F}}^d - (\bar{\mathbf{F}}^{d^T} \bar{\mathbf{F}}^d - \bar{\mathbf{F}}_0^{d^T} \bar{\mathbf{F}}_0^d) \end{aligned} \tag{4.30}$$

It should be noted that the deformation gradients following paths I and II are not related directly. However, the Green deformation tensors may be obtained following paths I or II. This is clearly indicated in equation (4.30) where one needs to remove $(\bar{\mathbf{F}}^{d^T} \bar{\mathbf{F}}^d)$ due to the fictitious removal of damage from the material. However, one needs to add the initial fictitiously removed deformation due to damage. Both equations (4.25) and (4.30) show that the Green deformation tensor can be expressed by either path I or path II, respectively. In addition, the effective Green deformation tensor of path II is the counterpart of the effective stress field and is defined as follows:

$$\bar{\mathbf{G}} = \bar{\mathbf{F}}^{e^T} \bar{\mathbf{F}}^e \tag{4.31}$$

For simplicity, one assumes that no initial damage exists in the initial undeformed body. Consequently, one obtains the following relation such that

$$\bar{\mathbf{F}}^{d^T} \bar{\mathbf{F}}_0^d = \mathbf{I} \tag{4.32}$$

$$\bar{\mathbf{F}}^e = \mathbf{F}^e \tag{4.33}$$

where $\mathbf{I}$ is a second order identity tensor. Using equations (4..31), (4.32), (4.33), one obtains the Green deformation tensor relating paths I and II as follows

$$\mathbf{G} = \mathbf{F}^{d^T} \bar{\mathbf{G}} \mathbf{F}^d \tag{4.34}$$

or

$$\mathbf{G} = \bar{\mathbf{F}}^{d^T} \bar{\mathbf{G}} \bar{\mathbf{F}}^d - (\bar{\mathbf{F}}^{d^T} \bar{\mathbf{F}}^d - \mathbf{I}) \tag{4.35}$$

From equation (4.35), one obtains the effective Green deformation tensor as follows

$$\begin{aligned}
\bar{\mathbf{G}} &= \bar{\mathbf{F}}^{d^{-T}} [\mathbf{G} + (\bar{\mathbf{F}}^{d^T} \bar{\mathbf{F}}^d - \mathbf{I})] \bar{\mathbf{F}}^{d^{-1}} \\
&= \bar{\mathbf{F}}^{d^{-T}} \mathbf{G} \bar{\mathbf{F}}^{d^{-1}} - \bar{\mathbf{F}}^{d^{-T}} \bar{\mathbf{F}}^{d^{-1}} + \mathbf{I}
\end{aligned} \tag{4.36}$$

Equating equations (4.34) and (4.35), one obtains the following relationship

$$\bar{\mathbf{F}}^{d^T} \bar{\mathbf{G}} \bar{\mathbf{F}}^d = \bar{\mathbf{F}}^{d^T} \bar{\mathbf{G}} \bar{\mathbf{F}}^d - \bar{\mathbf{F}}^{d^T} \bar{\mathbf{F}}^d + \mathbf{I} \tag{4.37}$$

The Green-Saint-Venant strain tensor termed here the strain tensor is defined as follows:

$$\epsilon = \tfrac{1}{2}(\mathbf{G} - \mathbf{I}) \tag{4.38}$$

The corresponding effective strain tensor of the counterpart of the effective stress is defined such that

$$\bar{\epsilon} = \tfrac{1}{2}(\bar{\mathbf{G}} - \mathbf{I}) \tag{4.39}$$

Substituting equation (4.36) into equation (4.39), one obtains the effective strain tensor in terms of the elastic-damage Green tensor and the fictitious effective damage gradient such that

$$\bar{\epsilon} = \tfrac{1}{2}\bar{\mathbf{F}}^{d^{-T}}(\mathbf{G} - \mathbf{I})\bar{\mathbf{F}}^{d^{-1}} \tag{4.40}$$

Finally one obtains the relation between $\bar{\epsilon}$ and ϵ using equations (4.38) and (4.40) such that

$$\bar{\epsilon} = \bar{\mathbf{F}}^{d^{-T}}\epsilon\bar{\mathbf{F}}^{d^{-1}} \tag{4.41}$$

Alternatively, the strain tensor is given by

$$\bar{\epsilon} = \bar{\mathbf{F}}^{d^{T}}\bar{\epsilon}\bar{\mathbf{F}}^{d} \tag{4.42}$$

The approach used here provides a relation between the effective strain and the damage elastic strain applicable also to finite strains and is not confined to small strains as in the case of the strain equivalence or the strain energy equivalence approach. Since the fictitious effective deformed configuration denoted by $\bar{\mathbf{C}}$, is obtained by removing the damage from the real deformed configuration denoted by $\mathbf{C}$, the fictitious effective deformed volume denoted by $\bar{\Omega}$ is similarly obtained as follows:

$$\begin{aligned}\bar{\Omega} &= \Omega - \Omega^{d} \\ &= \sqrt{(1 - \hat{\phi}_{1})(1 - \hat{\phi}_{2})(1 - \hat{\phi}_{3})}\,\Omega\end{aligned} \tag{4.43}$$

or

$$\Omega = \bar{J}^{d}\bar{\Omega} \tag{4.44}$$

where Ω is the deformed volume, Ω^{d} is the damage volume, and $\bar{J}^{d}$ is the Jacobian of the damage deformation. The Jacobian of the damage deformation is given by

$$\bar{J}^{d} = \frac{1}{\sqrt{(1 - \hat{\phi}_{1})(1 - \hat{\phi}_{2})(1 - \hat{\phi}_{3})}} \tag{4.45}$$

However, the Jacobian of the damage is defined such that

$$\begin{aligned}\bar{J}^{d} &= \sqrt{|\bar{\mathbf{G}}^{d}|} \\ &= \sqrt{|\bar{\mathbf{F}}^{d^{T}}\bar{\mathbf{F}}^{d}|} \\ &= \sqrt{|\bar{\mathbf{F}}^{d^{T}}||\bar{\mathbf{F}}^{d}|}\end{aligned} \tag{4.46}$$

The determinant of the matrix $[a]$ in equation (4.27) is given by

$$\begin{aligned}
||[a]|| &= ||[b]||^T ||[\hat{a}]|| ||[b]|| \\
&= ||[\hat{a}]|| \\
&= \frac{1}{\sqrt{(1 - \hat{\phi}_1)(1 - \hat{\phi}_2)(1 - \hat{\phi}_3)}}
\end{aligned} \tag{4.47}$$

Thus one assumes the following relation similar to equation (7) without loss of generality

$$\begin{aligned}
\bar{\sigma}_{ij} &= \hat{M}_{ikjl}\hat{\sigma}_{kl} \\
&= \hat{a}_{ik}\hat{a}_{jl}\hat{\sigma}_{kl} \\
&= \hat{\bar{F}}^d_{ik}\hat{\bar{F}}^d_{jl}\hat{\sigma}_{kl}
\end{aligned} \tag{4.48}$$

for stresses coinciding with the principal directions of damage. Consequently one obtains

$$\hat{\bar{F}}^d_{ij} = \hat{a}_{ij} \tag{4.49}$$

and

$$\bar{F}^d_{ij} = a_{ij} \tag{4.50}$$

Although the identity is established between $\bar{J}^d$ and $|a|$, this is not sufficient to demonstrate the validity of equation (4.49). Equation (4.49) is assumed here based on the physics of the geometrically symmetrized effective stress concept (Cordebois and Sidoroff, 1979; Voyiadjis and Park, 1996a,b). Equation (4.41) may now be expressed as follows:

$$\begin{aligned}
\bar{\epsilon}_{ij} &= a^{-1}_{ik}a^{-1}_{jl}\epsilon_{kl} \\
&= M^{-1}_{ijkl}\epsilon_{kl}
\end{aligned} \tag{4.51}$$

Similiarly, equation (4.42) can be written as follows

$$\begin{aligned}
\epsilon_{ij} &= a_{ik}a_{jl}\bar{\epsilon}_{kl} \\
&= M_{ijkl}\bar{\epsilon}_{kl}
\end{aligned} \tag{4.52}$$

The relations combining the strain of the elastic damage deformation and the effective strain in equations (4.51) and (4.52) indicate that these relationships are equivalent to those obtained using the hypothesis of energy equivalence (Cordebois and Sidoroff 1982).

4.6 Constitutive Equation of Elastic-Damage Behavior

The constitutive equation of the elastic-damage behavior is obtained by simple mapping rather than using the energy equivalence hypothesis. The constitutive equation in the effective configuration is given by

$$\bar{\sigma}_{ij} = \bar{E}_{ijkl}\bar{\epsilon}_{kl} \tag{4.53}$$

where $\bar{\mathbf{E}}$ is the elastic stiffness tensor of the material in the absence of damage. Substituting equations (4.7) and (4.51) into equation (4.53), the elastic-damage constitutive equation is obtained as follows:

$$\sigma_{ij} = N_{ikjl}\bar{E}_{klpq}N_{prqs}\epsilon_{rs}$$
$$= E_{ijrs}\epsilon_{rs} \tag{4.54}$$

The elastic damage stiffness is given as follows:

$$E_{ijrs} = N_{ikjl}\bar{E}_{klpq}N_{prqs} \tag{4.55}$$

where

$$N_{ikjl} = M_{ikjl}^{-1}$$
$$= a_{ik}^{-1}a_{jl}^{-1} \tag{4.56}$$

The elastic damage stiffness given by equation (4.55) is symmetric. This is in line with the classic (nonpolar) sense of continuum mechanics which is violated by using the hypothesis of strain equivalence. The damaged elastic stiffness is re-examined here using the fourth-order damage effective tensor $\mathbf{M}$ which is a function of the second-order damage tensor $\boldsymbol{\phi}$. The example of parallel cracks normal to the axis "1" is considered here. In this example the non-vanishing damage variables are the following

$$\phi_{11} = \hat{\phi}_1 \tag{4.57}$$

$$\phi_{22} = \hat{\phi}_2 \tag{4.58}$$

$$\phi_{33} = \hat{\phi}_3 \tag{4.59}$$

Since $\hat{\phi}_2 = \hat{\phi}_3 \ll \hat{\phi}_1$ one may assume for simplicity that $\phi_{22} = \phi_{33} = 0$. However, nonzeros for $\hat{\phi}_2$ and $\hat{\phi}_3$ are used here. The matrix form of the damaged elastic stiffness $\mathbf{E}$ is as follows:

$$
[E] = \begin{bmatrix}
\hat{\psi}_1\hat{\psi}_1\bar{E}_{1111} & \hat{\psi}_1\hat{\psi}_2\bar{E}_{1122} & \hat{\psi}_1\hat{\psi}_3\bar{E}_{1133} & 0 & 0 & 0 \\
\hat{\psi}_1\hat{\psi}_2\bar{E}_{1122} & \hat{\psi}_2\hat{\psi}_2\bar{E}_{2222} & \hat{\psi}_2\hat{\psi}_3\bar{E}_{2233} & 0 & 0 & 0 \\
\hat{\psi}_1\hat{\psi}_3\bar{E}_{1133} & \hat{\psi}_2\hat{\psi}_3\bar{E}_{2233} & \hat{\psi}_3\hat{\psi}_3\bar{E}_{3333} & 0 & 0 & 0 \\
0 & 0 & 0 & \hat{\psi}_1\hat{\psi}_2\bar{E}_{1212} & 0 & 0 \\
0 & 0 & 0 & 0 & \hat{\psi}_2\hat{\psi}_3\bar{E}_{2323} & 0 \\
0 & 0 & 0 & 0 & 0 & \hat{\psi}_1\hat{\psi}_3\bar{E}_{1313}
\end{bmatrix}
\tag{4.60}
$$

where $\hat{\psi}_i = (1 - \hat{\phi}_i)$. Alternatively, the matrix form of the damaged elastic compliance tensor $\mathbf{S}$ which is initially isotropic is given by

$$
[S] = \begin{bmatrix}
\frac{1}{\hat{\psi}_1\hat{\psi}_1\bar{E}} & \frac{-\bar{\nu}}{\hat{\psi}_1\hat{\psi}_2\bar{E}} & \frac{-\bar{\nu}}{\hat{\psi}_1\hat{\psi}_3\bar{E}} & 0 & 0 & 0 \\
\frac{-\bar{\nu}}{\hat{\psi}_1\hat{\psi}_2\bar{E}} & \frac{1}{\hat{\psi}_2\hat{\psi}_2\bar{E}} & \frac{-\bar{\nu}}{\hat{\psi}_2\hat{\psi}_3\bar{E}} & 0 & 0 & 0 \\
\frac{-\bar{\nu}}{\hat{\psi}_1\hat{\psi}_3\bar{E}} & \frac{-\bar{\nu}}{\hat{\psi}_2\hat{\psi}_3\bar{E}} & \frac{1}{\hat{\psi}_3\hat{\psi}_3\bar{E}} & 0 & 0 & 0 \\
0 & 0 & 0 & \frac{1}{\hat{\psi}_1\hat{\psi}_2\bar{G}} & 0 & 0 \\
0 & 0 & 0 & 0 & \frac{1}{\hat{\psi}_2\hat{\psi}_3\bar{G}} & 0 \\
0 & 0 & 0 & 0 & 0 & \frac{1}{\hat{\psi}_1\hat{\psi}_3\bar{G}}
\end{bmatrix}
\tag{4.61}
$$

where $\bar{\nu}$, $\bar{E}$, and $\bar{G}$ are the Poisson ratio, elastic and shear moduli, respectively. Obviously, the damaged stiffness or compliance tensor shown is for transversely isotropic material. The damaged material is limited to orthotropic behavior using the second-order damage tensor ϕ. However, the damaged material exhibits general orthotropy when the principal damage directions are not coincident with the current coordinates. In this case, all components are not zero. More detailed examples of damaged compliance for micro-crack distribution using scalar, second-order, fourth-order, and sixth-order damage tensors are availble in the recent work by Krajcinovic and Mastilovic (1995).

4.7 Conclusion

The fourth-order anisotropic damage effect tensor, $\mathbf{M}$, using the kinematic measure for damage expressed through the second-order damage tensor ϕ, is reviewed in this chapter in reference to the symmetrization of the effective stress tensor. This introduces a distinct kinematic measure of damage which is complementary to the deformation kinematic measure of strain. A thermodynamically consistent evolution equation for the damage tensor, ϕ together with a generalized thermodynamic force conjugate to the damage tensor is presented in the paper by Voyiadjis and Park (1995a,b). Voyiadjis and Venson (1995) quantified the physical values of the eigenvalues, $\hat{\phi}_k$ ($k = 1,2,3$), and the second-order damage tensor, ϕ, for the unidirectional fibrous composite by measuring the crack densities with the assumption that one of the eigen-directions of the damage tensor coincides with the fiber direction.

The fourth-order anisotropic damage effect tensor used here is obtained through the geometrical symmetrization of the effective stress (Cordebois and Sidoroff, 1979; Voyiadjis and Park, 1996a,b). This tensor is used here for the kinematic description of damage. The explicit representation of the fourth-order damage effect tensor is obtained with reference to the principal damage direction coordinate system.

The elastic-damage deformation for finite strain is also described in this chapter using the kinematics of damage. The kinematics of elastic-damage behavior proposed here allows one to obtain the strain tensor of the elastic-damage deformation without the use of either the hypotheses of energy equivalence or strain equivalence. The proposed approach provides a relation between the effective strain and the damage elastic strain applicable to finite strains and is not confined to small strains as in the case of the strain equivalence or strain energy equivalence approaches. This leads to a simpler derivation of the elastic-damage constitutive equation for the elastic-damage behavior without the use of the hypothesis of energy equivalence.

5

Anisotropic Damage Mechanics

5.1 Introduction

Following the first pioneering paper of Kachanov (1958) on damage mechanics, many different types of damage models have been introduced such as ductile fracture (Sidoroff, 1981; Cordebois and Sidoroff, 1979; Cordebois, 1983; Lee, et al., 1985; Chow and Wang, 1987, 1988), fatigue of metals (Lemaitre, 1971; Chaboche, 1974), creep and creep-fatigue problems (Leckie and Hayhurst, 1974; Hult, 1974; Lemaitre and Chaboche, 1975), and other types of damage (Lemaitre, 1984).

The concept of effective stress plays an important role in the definition of a suitable damage variable. The damage variable is usually defined to represent average material degradation that reflects all types of damage at the microscale level like nucleation and growth of voids, cavities, and other microcracks. The damage variable is assumed to be a second order tensor for the general case of anisotropic damage. The theory of anisotropic damage mechanics has been recently developed by Cordebois and Sidoroff (1979), Sidoroff (1981), Cordebois (1983), Chow and Wang (1987, 1988), Krajcinovic and Foneska (1981), Murakami and Ohno (1981), Murakami (1983), and Krajcinovic (1983), Voyiadjis and Kattan (1992,1999). The material in the following sections is based mainly on the work of Voyiadjis and Park (1996a,b).

A computational anisotropic damage model is developed in this chapter. An updated Lagrangian finite element formulation is used for the numerical solution of the problem. In this model, finite strains are used; however, small elastic strains are assumed in the formulation. In Figure 5.1 (Chow and Wang, 1987), it is clear that the assumption of small elastic strains is valid for this material behavior. It is also shown how the computational model developed here can be applied to problems of ductile fracture. The classical problem of a center-cracked thin plate subjected to in-plane tensile forces is analyzed using the proposed model. The results obtained are compared with the experimental results of Chow and Wang (1988).

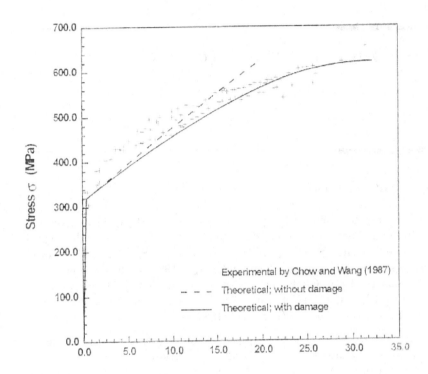

FIGURE 5.1
Uniaxial stress-strain curves. (Voyiadjis and Park, 1996b)

5.2   Anisotropic Damage Criterion

The anisotropy of damage hardening is considered here in order to obtain
a damage criterion for nonproportional loading. The damage criterion g is
expressed in terms of a tensorial hardening parameter $\mathbf{h}$ and the generalized
thermodynamic force $\mathbf{Y}$ conjugate to the damage tensor ϕ such that

$$g \equiv p_{ijkl}Y_{ij}Y_{kl} - 1 = 0 \tag{5.1}$$

where

$$p_{ijkl} = h_{ij}^{-1}h_{kl}^{-1} \tag{5.2}$$

Equation (5.1) is an isotropic function of $\mathbf{h}$ and $\mathbf{Y}$. The hardening tensor
$\mathbf{h}$ is expressed as

$$h_{ij} = (u_{ik})^{1/2} \phi_{kl} (u_{ij})^{1/2} + V_{ij} \tag{5.3}$$

$$[\mathbf{u}] = \begin{bmatrix} \lambda_1 q_1 \left(\frac{\kappa^D}{\lambda_1}\right)^{r_1} & 0 & 0 \\ 0 & \lambda_2 q_2 \left(\frac{\kappa^D}{\lambda_2}\right)^{r_2} & 0 \\ 0 & 0 & \lambda_3 q_3 \left(\frac{\kappa^D}{\lambda_3}\right)^{r_3} \end{bmatrix} \tag{5.4}$$

$$[\mathbf{V}] = \begin{bmatrix} \lambda_1 v_1^2 & 0 & 0 \\ 0 & \lambda_2 v_2^2 & 0 \\ 0 & 0 & \lambda_3 v_3^2 \end{bmatrix} \tag{5.5}$$

Tensors $\mathbf{u}$ and $\mathbf{V}$ are generalizations to orthotropic materials of the scalar forms of isotropic materials originally proposed by Stumvoll and Swoboda (1993). The quantities λ_1, λ_2, λ_3, v_1, v_2, v_3, r_1, r_2, r_3, q_1, q_2 and q_3 are material parameters obtained by matching the theory with experimental results. Parameters λ_1, λ_2 and λ_3 are related to the elasticity tensor $\mathbf{E}$, for an orthotropic material expressed by the 6×6 matrix

$$[\mathbf{E}] = \begin{bmatrix} (\lambda_1 + 2\mu_1) & \lambda_1 & \lambda_3 & 0 & 0 & 0 \\ \lambda_1 & (\lambda_2 + 2\mu_2) & \lambda_2 & 0 & 0 & 0 \\ \lambda_3 & \lambda_2 & (\lambda_3 + 2\mu_3) & 0 & 0 & 0 \\ 0 & 0 & 0 & \mu_1 & 0 & 0 \\ 0 & 0 & 0 & 0 & \mu_2 & 0 \\ 0 & 0 & 0 & 0 & 0 & \mu_3 \end{bmatrix} \tag{5.6}$$

such that

$$E_{1122} = \lambda_1$$

$$\frac{1}{2}(E_{1111} - E_{1122}) = \mu_1$$

$$E_{2233} = \lambda_2$$

$$\frac{1}{2}(E_{2222} - E_{2233}) = \mu_2$$

$$E_{1133} = \lambda_3$$

$$\frac{1}{2}(E_{3333} - E_{1133}) = \mu_2 \tag{5.7}$$

The parameters v_1, v_2 and v_3 in equation (5.5) define the initial threshold against damage for the orthotropic material. These parameters are obtained from the constraint that the onset of damage corresponds to the stress level at which the virgin material starts exhibiting nonlinearity. The scalar damage hardening parameter κ^D in equation (5.4) is given by

$$\kappa^D = \int_0^t -Y_{ij}\dot{\phi}_{ij}dt \qquad (5.8)$$

The damaging state is defined as any state that satisfies $g = 0$. Four different states are outlined below:

$$g < 0, \qquad\qquad\qquad\qquad (elastic - unloading) \qquad (5.9a)$$

$$g = 0, \frac{\partial g}{\partial \dot{Y}_{ij}}\dot{Y}_{ij} < 0 \qquad\qquad (elastic - unloading) \qquad (5.9b)$$

$$g = 0, \frac{\partial g}{\partial \dot{Y}_{ij}}\dot{Y}_{ij} = 0 \qquad\qquad (neutral\,loading) \qquad (5.9c)$$

$$g = 0, \frac{\partial g}{\partial \dot{Y}_{ij}}\dot{Y}_{ij} > 0 \qquad (loading\,for\,a\,damaging\,state) \qquad (5.9d)$$

An anisotropic damage criterion g is proposed here by expression (5.1). The corresponding loading conditions are given by equations (5.9). The damage criterion is expressed through the second order tensors **u** and **V** and the damage tensor ϕ.

5.3   Damage Tensor

In this section, the effective stress concept proposed by Kachanov (1958) and later generalized by Murakami (1988) is used. The initial undeformed and undamaged configuration of the body will be referred to as C_0 while $\bar{C}$ refers to the state of the body after it has only deformed without damage. The configuration of the body that is both deformed and damaged after a set of external agencies act on it is referred to as C (Voyiadjis and Kattan, 1992, 1999). A linear transformation is assumed between the Cauchy stress tensor σ, and the effective Cauchy stress tensor $\bar{\sigma}$ such that:

$$\bar{\sigma}_{ij} = M_{ijkl}\sigma_{kl} \qquad (5.10)$$

where **M** is the fourth-order damage effect tensor. Murakami (1988) has shown that **M** can be represented by 6×6 matrix as a function of $(\mathbf{I}_2 - \phi)$ such that

$$[\mathbf{M}] = [\mathbf{M}(\mathbf{I}_2 - \phi)] \qquad (5.11)$$

where ϕ is a symmetric second order tensor and $\mathbf{I}_2$ is the second-rank identity tensor. The effective Cauchy stress need not be symmetric or frame invariant under the given transformation. However, once the effective Cauchy stress is symmetrized, it can be shown that it satisfies the frame invariance principle (Voyiadjis and Kattan, 1992, 1999). The stress tensor **M** in conjunction with the matrix form of **M** may be represented in vector form as shown:

$$[\boldsymbol{\sigma}] = [\sigma_{11}, \sigma_{22}, \sigma_{33}, \sigma_{12}, \sigma_{23}, \sigma_{31}]^T \tag{5.12}$$

The symmetrized $\bar{\sigma}$ used here is given by (Lee, et al., 1986)

$$\bar{\sigma}_{ij} = \frac{1}{2}[\sigma_{ik}(\delta_{kj} - \phi_{ki})^{-1} + (\delta_{il} - \phi_{il})^{-1}\sigma_{lj}] \tag{5.13}$$

This stress, $\bar{\sigma}$, is frame independent. Making use of the symmetrization procedure given by equation (5.13), the (6×6) matrix form of tensor **M** is given by Voyiadjis and Kattan (1992, 1999) as follows:

$$[\mathbf{M}] = \frac{1}{2\nabla} \begin{bmatrix} 2\omega_{22}\omega_{33} - 2\phi_{23}^2 & 0 & 0 \\ 0 & 2\omega_{11}\omega_{33} - 2\phi_{13}^2 & 0 \\ 0 & 0 & 2\omega_{11}\omega_{22} - 2\phi_{12}^2 \\ \phi_{13}\phi_{23} + \phi_{12}\omega_{33} & \phi_{13}\phi_{23} + \phi_{12}\omega_{33} & 0 \\ 0 & \phi_{12}\phi_{13} + \phi_{23}\omega_{11} & \phi_{12}\phi_{13} + \phi_{23}\omega_{11} \\ \phi_{12}\phi_{23} + \phi_{13}\omega_{22} & 0 & \phi_{12}\phi_{23} + \phi_{13}\omega_{22} \end{bmatrix}$$

$$\begin{matrix} 2\phi_{13}\phi_{23} + 2\phi_{12}\omega_{33} & 0 \\ 2\phi_{13}\phi_{23} + 2\phi_{12}\omega_{33} & 2\phi_{12}\phi_{13} + 2\phi_{23}\omega_{11} \\ 0 & 2\phi_{12}\phi_{13} + 2\phi_{23}\omega_{11} \\ \omega_{22}\omega_{33} + \omega_{11}\omega_{33} - \phi_{23}^2 - \phi_{13}^2 & \phi_{12}\phi_{23} + \phi_{13}\omega_{22} \\ \phi_{12}\phi_{23} + \phi_{13}\omega_{22} & \omega_{11}\omega_{33} + \omega_{11}\omega_{22} - \phi_{13}^2 - \phi_{12}^2 \\ \phi_{12}\phi_{13} + \phi_{23}\omega_{11} & \phi_{13}\phi_{23} + \phi_{12}\omega_{33} \end{matrix}$$

$$\left.\begin{matrix} 2\phi_{12}\phi_{23} + 2\phi_{13}\omega_{22} \\ 0 \\ 2\phi_{12}\phi_{23} + 2\phi_{13}\omega_{22} \\ \phi_{12}\phi_{13} + \phi_{23}\omega_{11} \\ \phi_{13}\phi_{23} + \phi_{12}\omega_{33} \\ \omega_{22}\omega_{33} + \omega_{11}\omega_{22} - \phi_{23}^2 - \phi_{12}^2 \end{matrix}\right] \tag{5.14}$$

where ∇ is given by

$$\nabla = \omega_{11}\omega_{22}\omega_{33} - \phi_{23}^2\omega_{11} - \phi_{13}^2\omega_{22} - \phi_{12}^2\omega_{33} - 2\phi_{12}\phi_{23}\phi_{13} \tag{5.15}$$

The notation ω_{ij} is used to denote $\delta_{ij} - \phi_{ij}$. Using the principal damage values ϕ_1, ϕ_2, ϕ_3, then Equation (5.14) may be expressed as follows:

$$[\mathbf{M}]_{diag.} \begin{bmatrix} \frac{1}{1-\phi_1} & 0 & 0 & 0 & 0 & 0 \\ 0 & \frac{1}{1-\phi_2} & 0 & 0 & 0 & 0 \\ 0 & 0 & \frac{1}{1-\phi_3} & 0 & 0 & 0 \\ 0 & 0 & 0 & \frac{(1-\phi_2)+(1-\phi_1)}{2(1-\phi_2)(1-\phi_1)} & 0 & 0 \\ 0 & 0 & 0 & 0 & \frac{(1-\phi_3)+(1-\phi_2)}{2(1-\phi_3)(1-\phi_2)} & 0 \\ 0 & 0 & 0 & 0 & 0 & \frac{(1-\phi_3)+(1-\phi_1)}{2(1-\phi_3)(1-\phi_1)} \end{bmatrix}$$
$$(5.16)$$

However, in this chapter, the matrix representation of $\mathbf{M}$ for the case of plane stress will be used.

For the case of a thin plate subjected to a state of plane stress, it is assumed that the plate lies in the 1-2 plane under plane stress conditions. For plane stress, Equation (5.14) reduces to the following expression (Voyiadjis and Kattan, 1992, 1999)

$$\begin{Bmatrix} \bar{\sigma}_{11} \\ \bar{\sigma}_{22} \\ \bar{\sigma}_{12} \end{Bmatrix} = \begin{bmatrix} M_{11} & M_{12} & M_{13} \\ M_{21} & M_{22} & M_{23} \\ M_{31} & M_{32} & M_{33} \end{bmatrix} \begin{Bmatrix} \sigma_{11} \\ \sigma_{22} \\ \sigma_{12} \end{Bmatrix} \qquad (5.17)$$

where

$$M_{11} = \frac{(1-\phi_{22})}{\Delta} \qquad (5.18a)$$

$$M_{22} = \frac{(1-\phi_{11})}{\Delta} \qquad (5.18b)$$

$$M_{33} = \frac{(M_{11}+M_{22})}{2} \qquad (5.18c)$$

$$M_{12} = M_{21} = 0 \qquad (5.18d)$$

$$M_{13} = 2M_{31} = \frac{\phi_{12}}{\Delta} \qquad (5.18e)$$

$$M_{23} = 2M_{32} = M_{13} \qquad (5.18f)$$

$$\Delta = (1-\phi_{11})(1-\phi_{22}) - \phi_{12}^2 \qquad (5.18g)$$

The stresses and damage variables for the plane stress case are represented by 3×1 vectors such that

$$\{\sigma\} = [\sigma_{11}, \sigma_{22}, \sigma_{12}]^T \qquad (5.19a)$$

$$\{\phi\} = [\phi_{11}, \phi_{22}, \phi_{12}]^T \qquad (5.19b)$$

5.4 Damage Evolution

Although the two dissipative mechanisms of plasticity and damage influence each other, in this section, it is assumed that the energy dissipated due to plasticity and that due to damage are independent of each other. The power of dissipation Π is given by

$$\Pi = \Pi^D + \Pi^P \qquad (5.20)$$

where Π^P is the plastic dissipation and Π^D the corresponding damage dissipation. The plastic dissipation is given by

$$\Pi^P = \sigma_{ij}\dot{\epsilon}''_{ij} + \alpha_{ij}\dot{\beta}_{ij} + K^P\dot{\kappa}^P \qquad (5.21)$$

In this chapter, small elastic strains but finite inelastic deformations are assumed and the strain rate is assumed to be decomposed into an elastic component, $\dot{\epsilon}'_{ij}$ and a plastic component, $\dot{\epsilon}''_{ij}$, such that

$$\epsilon_{ij} = \epsilon'_{ij} + \epsilon''_{ij} \qquad (5.22)$$

From the experimental results shown in Figure 5.1 by Chow and Wang (1987), it is clearly indicated there that the elastic strains are extremely small (less than 0.2%) and therefore justifiably assumed to be small. In equation (5.21), the term $\alpha_{ij}\dot{\beta}_{ij}$ is associated with kinematic hardening and $K^P\dot{\kappa}^P$ with plastic isotropic hardening. The associated damage dissipation is given by

$$\Pi^D = Y_{ij}\dot{\phi}_{ij} + K^D\dot{\kappa}^D \qquad (5.23)$$

The fictitious undamaged material is characterized by the effective stress and effective strain. Since in the effective configuration, $\bar{C}$, the body has deformed without damage, hence the dissipation energy of the fictitious undamaged material is only composed of the plastic dissipation.

$$\bar{\Pi} = \bar{\Pi}^P \qquad (5.24)$$

and

$$\bar{\Pi} = \bar{\sigma}_{ij}\dot{\bar{\epsilon}}''_{ij} + \bar{\alpha}_{ij}\dot{\bar{\beta}}_{ij} + \bar{K}^P\dot{\bar{\kappa}}^P \qquad (5.25)$$

Since it is assumed that plastic yielding is independent of the damage process, consequently the plastic dissipation in the damaged material is equal to the corresponding plastic dissipation in the fictitious undamaged material. This leads to the following expression

$$\Pi^P = \bar{\Pi}^P \tag{5.26}$$

The following decomposition is further assumed:

$$\sigma_{ij}\dot{\epsilon}_{ij} = \bar{\sigma}_{ij}\dot{\bar{\epsilon}}_{ij}'' \tag{5.27a}$$

$$\alpha_{ij}\dot{\beta}_{ij} = \bar{\alpha}_{ij}\dot{\bar{\beta}}_{ij} \tag{5.27b}$$

$$K^P\dot{\kappa}^P = \bar{K}^P\dot{\bar{\kappa}}^P \tag{5.27c}$$

The assumptions imposed in order to obtain equation (5.27) from equation (5.26) is an attempt to simplify the problem in order to obtain a closed form expression for the stiffness matrix. Without these assumptions, the problem may not be solved. However, the good correlation between the experimental and numerical results provide a justification for this assumption. Making use of equations (5.27a) and (5.10) one obtains a transformation equation for the plastic strain rates such that

$$\dot{\bar{\epsilon}}_{ij}'' = M_{ijkl}^{-1}\dot{\epsilon}_{kl}'' \tag{5.28}$$

Using the method of the calculus of functions of several variables, one introduces two Lagrange multipliers $\dot{\Lambda}_1$ and $\dot{\Lambda}_2$ and forms the function Ω such that (Voyiadjis and Kattan, 1992, 1999)

$$\Omega = \Pi - \dot{\Lambda}_1 f - \dot{\Lambda}_2 g \tag{5.29}$$

where $f(\boldsymbol{\sigma}, \boldsymbol{\alpha})$ is the plastic yield function and $\boldsymbol{\alpha}$ is the backstress tensor. To extremize the function Ω, one uses the necessary conditions

$$\frac{\partial\Omega}{\partial\sigma_{ij}} = 0 \tag{5.30a}$$

and

$$\frac{\partial\Omega}{\partial Y_{ij}} = 0 \tag{5.30b}$$

which give the corresponding plastic strain rate and damage rate evolution equations, respectively:

$$\dot{\epsilon}_{ij}'' = \dot{\Lambda}_1 \frac{\partial f}{\partial\sigma_{ij}} \tag{5.31}$$

and

$$\dot{\phi}_{ij}'' = \dot{\Lambda}_2 \frac{\partial g}{\partial Y_{ij}} \tag{5.32}$$

Equation (5.32) gives the increment of damage from the potential g.
Using the consistency condition for damage,

$$\dot{g} = 0 \qquad (5.33)$$

which states that after an increment of damage the volume element again
must be in a damaging state, the quantity $\dot{\Lambda}_2$ is obtained such that

$$\dot{\Lambda}_2 = -\frac{\frac{\partial g}{\partial Y_{mn}}}{\frac{\partial g}{\partial \phi_{ij}}\frac{\partial g}{\partial Y_{ij}}}\dot{Y}_{mn} \qquad (5.34)$$

Substituting equation (5.34) into equation (5.32), one obtains

$$\dot{\phi}_{kl} = \Psi_{klij}\dot{Y}_{ij} \qquad (5.35)$$

where the fourth order tensor Ψ is given by

$$\Psi_{klij} = -\frac{\frac{\partial g}{\partial Y_{kl}}\frac{\partial g}{\partial Y_{ij}}}{\frac{\partial g}{\partial \phi_{mn}}\frac{\partial g}{\partial Y_{mn}}} \qquad (5.36)$$

The generalized thermodynamic force $\mathbf{Y}$, is assumed to be a function of
the elastic component of the strain tensor ϵ', and the damage tensor ϕ, or
the stress σ and ϕ

$$\mathbf{Y} = \mathbf{Y}(\epsilon', \phi) \qquad or \qquad \mathbf{Y} = \mathbf{Y}(\sigma, \phi) \qquad (5.37)$$

The evolution equation for $\mathbf{Y}$ may be expressed as follows:

$$\dot{Y}_{ij} = \frac{\partial Y_{ij}}{\partial \sigma_{mn}}\dot{\sigma}_{mn} + \frac{\partial Y_{ij}}{\partial \phi_{pq}}\dot{\phi}_{pq} \qquad (5.38)$$

Substituting for $\dot{\mathbf{Y}}$ from equation (5.38) into equation (5.35) one obtains

$$\dot{\phi}_{kl} = \Psi_{klij}\left(\frac{\partial Y_{ij}}{\partial \sigma_{mn}}\dot{\sigma}_{mn} + \frac{\partial Y_{ij}}{\partial \phi_{pq}}\dot{\phi}_{pq}\right) \qquad (5.39)$$

Equation (5.39) may be further reduced to

$$L_{ijkl}\dot{\phi}_{kl} = \Psi_{ijrs}\frac{\partial Y_{rs}}{\partial \sigma_{mn}}\dot{\sigma}_{mn} \qquad (5.40)$$

where

$$L_{ijkl} = \frac{1}{2}(\delta_{ik}\delta_{jl} + \delta_{il}\delta_{jk}) - \Psi_{ijmn}\frac{\partial Y_{mn}}{\partial \phi_{kl}} \qquad (5.41)$$

Equation (5.40) is used to solve for $\dot{\phi}$ such that

$$\dot{\phi}_{kl} = \left(L_{ijkl}^{-1} \Psi_{ijrs} \frac{\partial Y_{rs}}{\partial \sigma_{mn}} \right) \dot{\sigma}_{mn} \qquad (5.42a)$$

or

$$\dot{\phi}_{kl} = T_{klmn} \dot{\sigma}_{mn} \qquad (5.42b)$$

The thermodynamic force associated with damage is obtained using the enthalpy of the damaged material where

$$V(\sigma, \phi) = \sigma_{mn} \epsilon'_{mn} - W \qquad (5.43)$$

or

$$V = \frac{1}{2} \sigma_{mn} E_{mnkl}^{-1}(\phi) \sigma_{kl} \qquad (5.44)$$

and W is the specific energy. In equation (5.44), $\mathbf{E}$ is the damaged elasticity tensor. The thermodynamic force is given by

$$Y_{ij} = -\frac{\partial V}{\partial \phi_{ij}} \qquad (5.45a)$$

or

$$Y_{ij} = \frac{-\partial V}{\partial M_{abcd}} \frac{\partial M_{abcd}}{\partial \phi_{ij}} \qquad (5.45b)$$

Using the energy equivalence principle, one obtains a relation between the damaged elasticity tensor $\mathbf{E}$ and the effective, undamaged, elasticity tensor $\bar{\mathbf{E}}$ such that (Voyiadjis and Kattan, 1992, 1999)

$$E_{mnkl}^{-1}(\phi) = M_{uvmn}(\phi) \bar{E}_{uvpq}^{-1} M_{pqkl}(\phi) \qquad (5.46)$$

Making use of equations (5.44) and (5.45), the thermodynamic force is given explicitly as follows:

$$Y_{ij} = \frac{-1}{2} (\sigma_{cd} \bar{E}_{abpq}^{-1} M_{pqkl} \sigma_{kl} + \sigma_{kl} M_{pqkl} \bar{E}_{pqab}^{-1} \sigma_{cd}) \frac{\partial M_{abcd}}{\partial \sigma_{ij}} \qquad (5.47)$$

5.5 Constitutive Model

In this section, the elasto-plastic stiffness matrix that involves damage effects is derived. Rate-dependent effects are neglected and isothermal conditions are assumed.

The incremental form of equation (5.10) may be expressed as

$$\dot{\bar{\sigma}}_{ij} = \dot{M}_{ijkl}\sigma_{kl} + M_{ijkl}\dot{\sigma}_{kl} \tag{5.48}$$

Making use of the additive decomposition of the effective strain rate and equation (5.28), one obtains

$$\dot{\bar{\epsilon}}_{ij} = \dot{M}_{ijkl}^{-1}\epsilon'_{kl} + M_{ijkl}^{-1}\dot{\epsilon}_{kl} \tag{5.49}$$

The rates of the damage effect tensor and its inverse may be expressed as follows by making use of equation (5.42b)

$$\dot{M}_{ijkl} = \frac{\partial M_{ijkl}}{\partial \phi_{pq}}T_{pqmn}\dot{\sigma}_{mn} \tag{5.50a}$$

$$= Q_{ijklmn}\dot{\sigma}_{mn} \tag{5.50b}$$

and

$$\dot{M}_{ijkl}^{-1} = \frac{\partial M_{ijkl}^{-1}}{\partial \phi_{pq}}T_{pqmn}\dot{\sigma}_{mn} \tag{5.51a}$$

$$= R_{ijklmn}\dot{\sigma}_{mn} \tag{5.51b}$$

In order to obtain the elasto-plastic stiffness matrix that involves damage effects, we start with the elasto-plastic stiffness relation in the undamaged configuration given by Voyiadjis and Kattan (1992, 1999)

$$\dot{\bar{\sigma}}_{ij} = \bar{D}_{ijkl}\dot{\bar{\epsilon}}_{kl} \tag{5.52}$$

The elasto-plastic effective stiffness $\bar{D}$ is given by Voyiadjis and Kattan (1992, 1999).

$$\bar{D}_{klmn} = \bar{E}_{klmn} - \frac{1}{Q}\frac{\partial \bar{f}}{\partial \bar{\sigma}_{pq}}\bar{E}_{pqmn}\bar{E}_{klij}\frac{\partial \bar{f}}{\partial \bar{\sigma}_{ij}} \tag{5.53}$$

$$\bar{Q} = \frac{\partial \bar{f}}{\partial \sigma_{ab}}\bar{E}_{abcd}\frac{\partial \bar{f}}{\partial \bar{\sigma}_{cd}} - \frac{\partial \bar{f}}{\partial \kappa^p}\bar{\sigma}_{pq}\frac{\partial \bar{f}}{\partial \bar{\sigma}_{pq}} - \frac{\partial \bar{f}}{\partial \bar{\sigma}_{ef}}(\bar{\sigma}_{ef} - \bar{\alpha}_{ef})b\frac{\frac{\partial \bar{f}}{\partial \bar{\sigma}_{ij}}\frac{\partial \bar{f}}{\partial \bar{\sigma}_{ij}}}{(\bar{\sigma}_{gh} - \bar{\alpha}_{gh})\frac{\partial \bar{f}}{\partial \bar{\sigma}_{gh}}} \tag{5.54}$$

This is valid for a yield function of the form

$$f = \frac{3}{2}(\bar{\sigma}_{kl} - \bar{\alpha}_{kl})(\bar{\sigma}_{kl} - \bar{\alpha}_{kl}) - \bar{\sigma}_0^2 - c\bar{\kappa}^P = 0 \tag{5.55}$$

The plastic work $\bar{\kappa}^P$ is a scalar function and its evolution in the configuration $\bar{C}$ is taken to be in the form

$$\dot{\bar{\kappa}}^P = (\dot{\bar{\epsilon}}_{ij}'' \dot{\bar{\epsilon}}_{ij}'')^{1/2} \tag{5.56}$$

A Prager-Ziegler kinematic hardening evolution law is used such that

$$\dot{\bar{\alpha}}_{ij} = \dot{\bar{\mu}}(\bar{\sigma}_{ij} - \bar{\alpha}_{ij}) \tag{5.57}$$

Making use of equations (5.52), (5.49), (5.50b) and (5.51b), one obtains the resulting elastoplastic stiffness relation in the damaged configuration as follows:

$$\dot{\sigma}_{mn} = D_{mnpq}\dot{\epsilon}_{pq} \tag{5.58}$$

where

$$D_{mnpq} = O_{mnij}^{-1} \bar{D}_{ijkl} M_{klpq}^{-1} \tag{5.59}$$

and

$$O_{ijkl} = Q_{ijmnkl}\sigma_{mn} + M_{ijkl} - \bar{D}_{ijmn}R_{mnpqkl}E_{pqab}^{-1}\sigma_{ab} \tag{5.60}$$

It is interesting to note that for the case of no damage, tensors **Q** and **R** reduce to zero and **M** becomes a fourth order identity tensor. In this case, equation (5.58) reduces to equation (5.52).

5.6 Uniaxial Tension Analysis

In order to obtain the damage parameters, uniaxial tension analysis for the aluminum alloy 2024-T3 is simulated numerically and compared with experimental results (Chow and Wang, 1987, 1988). The damage parameters q_l, q_2, q_3, r_l, r_2 and r_3 are selected such that the computed results present a best fit of the experimental data. The stress-strain curves shown in Figure 5.1 show good agreement with the experimental data by Chow and Wang (1987). In addition to the stress-strain curve, damage evolutions with respect to the stress and strain are shown in Figure 5.2 and 5.3, respectively. It is clearly shown that the growth of damage with stress accelerates rapidly near failure. This is in accord with the experiments of Chow and Wang (1987, 1988) whereby a substantial increase in the crack density is instrumental in forming links between the cracks that leads to failure.

5.7 Finite Element Implementation

The model is implemented numerically using an updated Lagrangian finite element method. The basic assumptions and equations for the finite element formulations have been presented by Kattan and Voyiadjis (1990) and Voyiadjis and Kattan (1990). The final incremental equilibrium equations in the updated Lagrangian description are given by (Kattan and Voyiadjis, 1990)

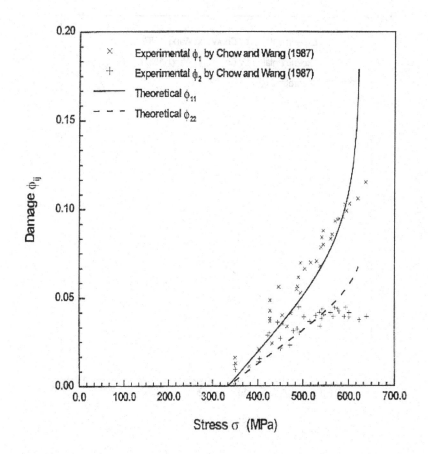

FIGURE 5.2
Uniaxial stress-damage curves. (Voyiadjis and Park, 1996b)

$$([K] + [K]^\sigma + [K]^{(NC)})\{dv\} = \{dP\} \qquad (5.61)$$

where $\{dv\}$ is the unknown incremental vector for the nodal displacements and $\{dP\}$ is the corresponding incremental vector for the nodal forces. $[K]$ is the symmetric "large displacement" matrix, $[K]^\sigma$ is the symmetric "initial stress" matrix, and $[K]^{(NC)}$ is the nonsymmetric "displacement dependent load" matrix. These matrices are given by

$$K_{ab} = \int \int \int \frac{\partial N_{ia}}{\partial x_j} D_{ijkl} \frac{\partial N_{kb}}{\partial x_l} dV \qquad (5.62a)$$

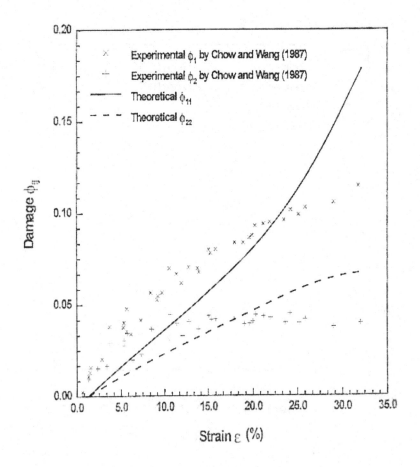

FIGURE 5.3

Uniaxial strain-damage curves. (Voyiadjis and Park, 1996b)

$$K_{ab}^{(\sigma)} = \int \int \int \frac{\partial N_{ka}}{\partial x_i} \sigma_{ij} \frac{\partial N_{kb}}{\partial x_j} dV \qquad (5.62b)$$

and

$$K_{ab}^{(NC)} = \int \int \int \varrho \frac{\partial p_i}{\partial x_j} N_{jb} N_{ia} dV + \int \int T_{ib} N_{ia} dA \qquad (5.62c)$$

where N_{ij} are the shape functions, T_{ib} are defined by the following relation (Zienkiewicz, 1977).

$$\frac{\partial t_i}{\partial x_j} u_j = T_{ib} q_b \qquad (5.63)$$

and q_b are the incremental nodal displacements. The above equations are applicable to the model presented provided the appropriate elasto-plastic, damage stiffness matrix $\mathbf{D}$ is substituted. The incremental vector for the nodal forces is given as follows:

$$dP_\alpha = \int \int \int \varrho(dp_i) N_{ia} dV + \int \int (dt_i) N_{ia} dA \qquad (5.64)$$

The damage model used in this work is successfully implemented in the fmite element program DNA (Damage Nonlinear Analysis). A convergence criteria is used to terminate the equilibrium iteration at each load increment. At the end of each iteration, the solution obtained is checked using the internal energy criterion $\Delta W_n^{(i)}$ as follows:

$$\{\Delta u\}^{(i)T} (^{m+1}\{R\} - ^{m+1}\{F\}^{(i)}) \leq \epsilon_E \{\Delta u\}^{(i)T} (^{m+1}\{R\} - ^m\{F\}) \quad (5.65)$$

where $\{\Delta u\}^{(i)}$ is the displacement increment obtained in the ith iteration, T is used to denote the transpose of a vector, and $^m\{R\}$ and $^m\{F\}$ are the internal force and external force vectors in the mth increment, respectively. The left-hand side of the inequality represents the work done by the out-of-balance force on the displacement increment, and the right-hand side is the initial value of the same work. In equation (5.63), ϵ_E is a prescribed tolerance for internal energy.

5.8 Center-Cracked Thin Plate Under In-Plane Tensile Forces

The proposed model is primarily derived to solve problems in ductile fracture. As an example, the problem of crack initiation in a center-cracked

TABLE 5.1
Material properties and parameters (aluminum alloy
2024-T3).

Modulus	73, 087 MPa
Poisson's ration	0.3
Yielding stress	330 MPa
Kinematic hardening parameter	$b = 275.8$ MPa
Isotropic hardening parameter	$c = 792.9$ MPa
Damage parameters	$r_1 = r_2 = r_3 = 0.64$
	$q_1 = q_2 = q_3 = 0.53$
	$v_1 = v_2 = v_3 = 0.006$

thin plate, shown in Figure 5.4, that is subjected to in-plane tension is
analyzed. The material used is aluminum alloy 2024 T3. All material
properties and parameters are shown in Table 5.1. Since the thickness of
the plate ($t = 3.175$ mm) is small compared with the other dimension, a
state of plane stress is assumed.

Due to symmetry in geometry and loading, only one-quarter of the plate
is discretized by finite elements as shown in Figure 5.5. The eight-node
quadrilateral isoparametric element is used in this finite element analysis.
It is noticed that a large number of regular elements are used around the
crack tip in order to avoid the special (singularity) elements at that point
(Henshell and Shaw, 1975; Barsoum, 1976). Consequently, a total of 381
elements and 1228 nodes are used. The load is incremented with uniform
load increments of 5 MPa until the principal damage value ϕ_p reaches 1.0
($\phi_p \leq 1.0$). The principal damage variable ϕ_p is given by:

$$\phi_p = \frac{\phi_{xx} + \phi_{yy}}{2} + \sqrt{\left(\frac{\phi_{xx} + \phi_{yy}}{2}\right)^2 + \phi_{xy}^2} \qquad (5.66)$$

Consequently, material failure, ϕ_c, occurs when $\phi_c = 0.38$, that is
$\phi_p = 0.38$. The principal damage value is monitored at each load increment since it is used to determine the onset of macro-crack initiation.
The onset of macro-crack initiation occurs at the front of crack tip when
the final load of 265 MPa is reached. This is compared with the crack
initiation load of 263.6 MPa obtained from experiments done by Chow and
Wang (1988). The contours of stresses σ_{xx}, σ_{yy} and σ_{xy}, for the quarter
plate and around the crack tip for the load 260 MPa are shown in Fig-
ures 5.6 and 5.7, respectively. It is clear from Figure 5.6 and 5.7 that high
stresses are localized around the crack tip with a maximum $\sigma_{yy} = 622$ MPa.

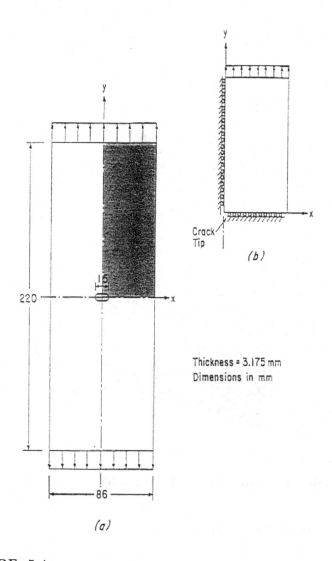

(a)

FIGURE 5.4

(a) This plate with a center crack. (b) Quarter of plate to be discretized by finite elements.(Voyiadjis and Park, 1996b)

Figure 5.8 shows the contours for the damage variables ϕ_{xx}, ϕ_{yy} and ϕ_{xy}. The maximum damage value of 0.28 (ϕ_{yy}) occurs near the tip of the crack. All the damage values stabilize when the crack initiation load of 260 MPa is reached. Once this load is attained, damage increases without further

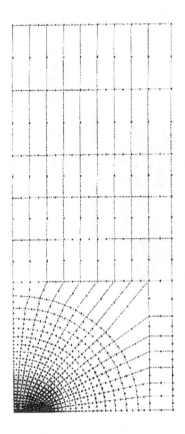

FIGURE 5.5
Finite element mesh. (Voyiadjis and Park, 1996b)

increase in load, thus initiating macrocracks.

(a) (b) (c)

FIGURE 5.6

Contour lines for stresses for the quarter plate at a load of P = 260 MPa
(a) Stress σ_{xx} contours in MPa. (b) Stress σ_{yy} contours in MPa. (c) Stress
σ_{xy} contours in MPa. (Voyiadjis and Park, 1996b)

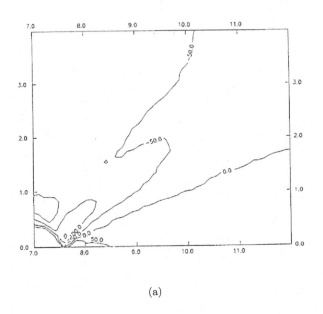

(a)

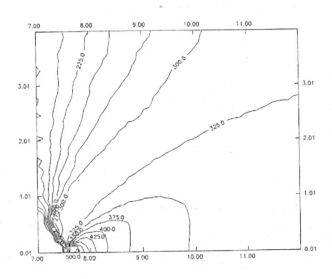

(b)

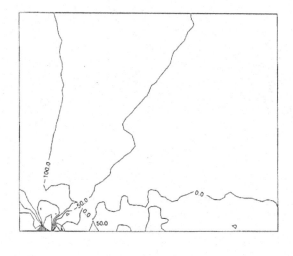

(c)

FIGURE 5.7

Contour lines for the stresses around the crack at a load of P = 260MPa:
(a) stress σ_{xx} contours in MPa, (b) stress σ_{yy} contours in MPa, (c) stress
σ_{xy} contours in MPa. (Voyiadjis and Park, 1996b)

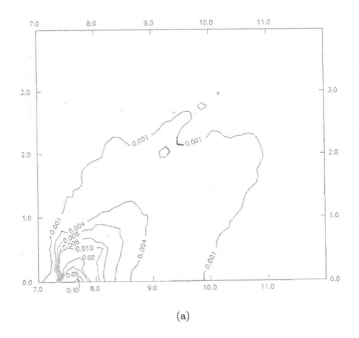

(a)

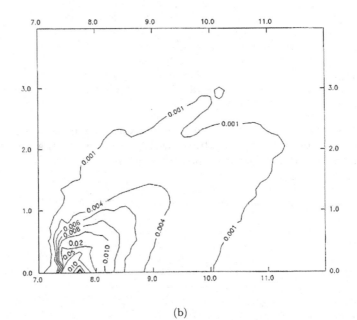

(b)

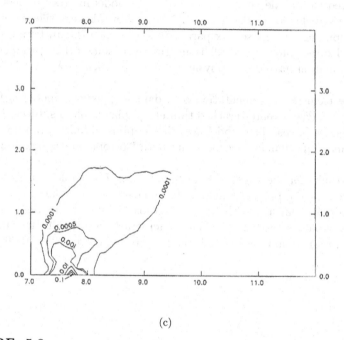

(c)

FIGURE 5.8

Contour lines for the stresses around the crack at a load of P = 260MPa:
(a) stress σ_{xx} contours in MPa, (b) stress σ_{yy} contours in MPa, (c) stress
σ_{xy} contours in MPa. (Voyiadjis and Park, 1996b)

5.9   Conclusion

An anisotropic damage model is proposed for the characterization of the degradation of the material properties as well as for the onset of macro-crack initiation in metals. The evolution and growth of damage using the proposed theory compares well with the experimental data from the uniaxial tension test given by Chow and Wang (1987). The authors have attempted to provide an anisotropic damage model in order to accurately predict the behavior of the material. The use of this versatile and general anisotropic model imposes six parameters for which the authors have not obtained direct physical correlation. However, some other parameters in the formulation have direct physical correlation such as λ_1, λ_2, λ_3, v_1, v_2 and v_3.

Finite element implementation of a damage-plasticity model is formulated to analyze a center-cracked thin elasto-plastic plate subjected to in-plane tensile forces. The model uses the coupling of finite strain plasticity and damage mechanics incorporating both isotropic and kinematic hardening.

The results for the stresses, strains and damage variables are shown and compared for the proposed model. It is shown that the damage variables increase monotonically with the applied load. The critical values for the damage variables needed to initiate macrocracks are within the range 0.2 to 0.8. This is in agreement with the work of Lemaitre (1984, 1986).

6

Plasticity Theory

The nonlinear behavior of metals when subjected to monotonic, and cyclic non-proportional loading is modeled using the proposed hardening rule. The plasticity model is based on the Chaboche (1989, 1991) and Voyiadjis and Sivakumar (1991, 1994) models incorporating the bounding surface concept. The evolution of the backstress is governed by the deviatoric stress rate direction, the plastic strain rate, the backstress, and the proximity of the yield surface from the bounding surface. In order to ensure uniqueness of the solution, nesting of the yield surface with the bounding surface is ensured. The prediction of the model in uniaxial cyclic loading is compared with the experimental results obtained by Chaboche (1989, 1991). The behavior of the model in multiaxial stress space is tested by comparing it with the experimental results in axial and torsional loadings performed by Shiratori, et al. (1979) for different stress trajectories. The amount of hardening of the material is tested for different complex stress paths. The model gives a very satisfactory result under uniaxial, cyclic and biaxial non-proportional loadings. Ratchetting is also illustrated using a non-proportional loading history.

6.1 Introduction

Voyiadjis and Sivakumar (1991, 1994) proposed a robust kinematic hardening model that ensures the nesting of the yield and bounding surfaces. This is accomplished by defining a kinematic hardening rule that blends the deviatoric stress rate rule for the movement of the yield surface with that of the Tseng and Lee rule (1983). This approach satisfies both the experimental observations made by Phillips, et al. (1974), and the nesting of the yield surface with the bounding surface. A general expression for the plastic modulus is also proposed in this chapter. The expressions proposed by McDowell (1987) and Dafalias (1981) may be obtained as a special case of that proposed by Voyiadjis and Sivakumar (1991, 1994). An additional parameter is also introduced in this work that reflects the dependence of the plastic modulus on the angle between the deviatoric stress rate tensor and the tensor joining the center of the bounding surface and the yield surface. The movement of the center of the yield surface is along a curved path in this case and not along a straight line path as predicted by Tseng and Lee (1983). The Voyiadjis and Sivakumar model (1994) predicts the yielding and ratchetting behavior of the material under cyclic loading quite well as compared with the experimental results of Shiratori, et al. (1979). However, the effective plastic strains are overpredicted at low values of the plastic strains.

The objective of this chapter is to develop a constitutive model that can describe the plastic response of class M materials under complex loading conditions specifically for hardening effects. An attempt is made to formulate a new kinematic hardening rule that can predict the response of the material under both monotonic and cyclic loadings. The model is based on both the Chaboche model (1989, 1991) and the Voyiadjis-Sivakumar model (1991, 1994). A new term involving the stress rate is incorporated in the evolution equation of the backstress along with a coefficient that depends on the proximity of the yield surface from the bounding surface. The prediction of the model in uniaxial monotonic and cyclic loading is compared with the experimental results obtained by Chaboche (1989, 1991). The behavior of the model in multiaxial stress space is tested by comparing it with the experimental results in axial and torsional loadings performed by Shiratori, et al. (1979) for different stress trajectories.

6.2   Theoretical Formulation

The model is developed in the multidimensional stress space. The yield surface is of a von Mises type given as follows:

$$f \equiv \frac{3}{2}(\mathbf{s} : \mathbf{s}) - k^2 = 0 \tag{6.1}$$

where

$$\mathbf{s} = \boldsymbol{\tau} - \mathbf{X} \tag{6.2}$$

The yield surface is expressed in terms of deviatoric components, $\boldsymbol{\tau}$, of the Cauchy stress tensor, $\boldsymbol{\sigma}$, and the backstress tensor $\mathbf{X}$. The tensorial operation ":" on second order tensors implies the the following:

$$\mathbf{s} : \mathbf{s} = s_{ij}s_{ij} \tag{6.3}$$

In equation (6.1), k is the radius of the yield surface. The initial size of the yield surface is given by k_0, defining the initial yield strength of the material in the uniaxial tension test. It should be noted that for class M materials, the hydrostatic component of the stress tensor is assumed to have no effect on the plastic deformation. An additive decomposition of the strain rate tensor, $\dot{\epsilon}$, is assumed such that

$$\dot{\epsilon} = \dot{\epsilon}' + \dot{\epsilon}'' \tag{6.4}$$

where $\dot{\epsilon}'$ is the elastic component of the strain rate tensor, and $\dot{\epsilon}''$ the corresponding plastic strain rate component. An associative flow rule is assumed in this formulation such that

$$\dot{\epsilon}'' = \dot{\Lambda}\frac{\partial f}{\partial \boldsymbol{\sigma}} \tag{6.5}$$

where $\dot{\Lambda}$ is a Lagrangian multiplier in conjunction with the flow rule.

Armstrong and Frederick (1966) proposed a kinematic hardening rule in which the evolution equation for the backstress, $\dot{\mathbf{X}}$, is given as follows:

$$\dot{\mathbf{X}} = \frac{2}{3}C\dot{\epsilon}'' - \gamma\mathbf{X}\dot{\mathbf{p}} \tag{6.6}$$

where C and γ are constants and p is the accumulated plastic strain rate,

$$\dot{p} = \sqrt{\frac{2}{3}\dot{\epsilon}''_{ij}\dot{\epsilon}''_{ij}} \tag{6.7}$$

Chaboche (1989, 1991) proposed a model based on Armstrong and Frederick (1966) in which the evolution equation of the backstress is decomposed into several components. The backstress is calculated as the sum of $\mathbf{X}^{(i)}$ components such that

$$\mathbf{X} = \sum_{i=1}^{N}\mathbf{X}^{(i)} \tag{6.8}$$

where

$$\dot{\mathbf{X}}^{(i)} = \frac{2}{3}C^{(i)}\dot{\epsilon}'' - \gamma^{(i)}\mathbf{X}^{(i)}\dot{\mathbf{p}} \tag{6.9}$$

and the summation in equation (6.8) varies depending on the model type. Chaboche (1989, 1991) proposed different models, namely, the short range kinematic model (NLK), the long rate kinematic model with Prager linear-kinematic rule (LK), the non-linear Prager rule (NLP2), the model with a modified recall term with a power function called MILL, and the model with a modified recall term with a threshold. (NLK-T). Each model has different values for the constants $C^{(i)}$ and $\gamma^{(i)}$. Chaboche (1991) showed that the NLK-T model best predicts the experimental results. This model introduces a nonlinearity in the recall term through a threshold term $\boldsymbol{X}$. The model is used to simulate the experiments performed on type 316 stainless steel. The model gives good correlation with experimental results for uniaxial monotonic, normal cyclic, and ratchetting conditions. However, when a small cycle is incorporated during a transient in a large cycle, the small cycle loop does not close the loop as it is anticipated from experimental observations. The closure of the small loop is not completed. The modeling of such behavior needs a discrete memory scheme to reproduce the above result.

Isotropic hardening is also incorporated in the proposed model. The isotropic hardening rule was proposed initially by Chaboche (1989) and generalized by Ohno (1982). The model introduces a surface of nonhardening in the plastic strain space. Ohno and Wang (1991, 1993) showed in their work that the formulation as given by Chaboche (1989) can be shown to be a multi-surface model. They show that the multi-surfaces generated are nested and obey an Mroz-type (1967, 1969) translation rule. They also show that for each $\mathbf{X}$ there exists a bounding surface in the $\mathbf{X}$ space and $\dot{\mathbf{X}}$ depends on the radius of its bounding surface, its proximity from the bounding surface, and the plastic strain rate $\dot{\epsilon}''$. The yield surface moves with the bounding surfaces with all the surfaces nesting each other at the point of loading with the center of each surface lying on a straight line. However, a general single bounding surface for all the components of $\mathbf{X}$ is not used which allows more flexibility and simplicity in accurately predicting any combination of monotonic and cyclic loads.

In the present chapter, a new term is added to the Armstrong-Frederick (1966) kinematic hardening rule to allow for the motion of the yield surface to be influenced by the direction of the stress rate as indicated by Phillips, et al. (1974). The kinematic hardening rule as given by Voyiadjis and Basuroychowdhury (1998) is as follows:

$$\dot{\mathbf{X}}^{(i)} = \frac{2}{3} C^{(i)} \dot{\epsilon}'' - \gamma^{(i)} \dot{\mathbf{X}}^{(i)} \dot{p} + \frac{1\delta}{m} \beta^{(i)} \qquad (6.10a)$$

and

$$\dot{\mathbf{X}} = \sum_{i=1}^{4} \dot{\mathbf{X}}^{(i)} \qquad (6.10b)$$

where $\beta^{(i)}$ is a material parameter and $\mathbf{l}$ is the direction of the stress rate.

In equation (6.10a), m is the length of the chord of the bounding surface along the direction of loading as indicated in Figure 6.1. The proximity of the yield surface from the bounding surface is given by δ which is the

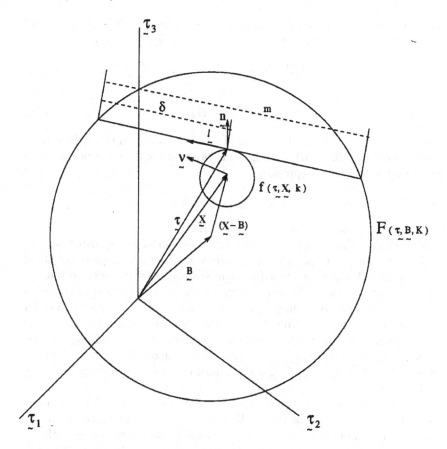

FIGURE 6.1
Proposed two surface kinematic hardening model.(Voyiadjis and Basuroy-chowdury, 1998)

distance of the stress point from the bounding surface in the direction of the stress rate tensor. Referring to Figure 6.1, δ and m are derived as follows:

$$\| \boldsymbol{\tau} - \mathbf{B} \| = \sqrt{(\boldsymbol{\tau} - \mathbf{B}) : (\boldsymbol{\tau} - \mathbf{B})} \qquad (6.11)$$

The measure of the projection of $\boldsymbol{\tau} - \mathbf{B}$ along the direction of the stress rate is given by the following expression

$$(\boldsymbol{\tau} - \mathbf{B}) : \mathbf{l} = (\tau_{ij} - B_{ij})l_{ij} \qquad (6.12)$$

where **B** is the backstress tensor defining the center of the bounding surface. The lengths m and δ are given by the following expressions:

$$m = 2\sqrt{K^2 - (\| \boldsymbol{\tau} - \mathbf{B} \|)^2 + [(\boldsymbol{\tau} - \mathbf{B}) : l]^2} \qquad (6.13)$$

$$\delta = \frac{m}{2} - (\boldsymbol{\tau} - \mathbf{B}) : l \qquad (6.14)$$

Equations (6.13) and (6.14) are valid until any point on the yield surface touches the bounding surface. The constraint for such a condition may be given by the following expression:

$$\| \mathbf{X} - \mathbf{B} \| + \mathbf{k} \le \mathbf{K} \qquad (6.15)$$

where K is the radius of the bounding surface.

Referring to Figure 6.1, the motion of yield surface is adjusted so that it does not intersect the bounding surface on contact. Equations (6.10) are valid for the movement of the yield surface until the surface touches the bounding surface. As soon as any point on the yield surface comes in contact with the bounding surface, the yield surface is rotated in order to allow the loading point on the yield surface to touch the bounding surface without allowing the yield surface to intersect the bounding surface. Thereafter, the movement of the center of the yield surface is governed by the movement of the bounding surface until the two surfaces separate from each other.

Once the yield surface touches the bounding surface, the yield surface is rotated in such a way that the loading point on the yield surface comes in contact with the bounding surface. The yield surface then moves with the bounding surface and, consequently, the movement of the yield surface is constrained by the movement of the bounding surface. The movement of the center of the yield surface during the process of rotation of the yield surface until the loading point on the yield surface comes in contact with the bounding surface is shown in Figure 6.2. Referring to Figure 6.2, **p** is a unit tensor along the direction of the backstress when the point of loading on the yield surface touches the bounding surface. $\mathbf{X^N}$ is the location of the center of the yield surface at contact with the bounding surface when the point of loading is on the bounding surface. $\mathbf{X^P}$ is the location of the center of the yield surface when initial contact with the bounding surface has occurred but the point of loading is not on the bounding surface. In Figure 6.2, L is the distance AB, and γ is the angle between the stress tensor $(\boldsymbol{\tau} - \mathbf{B})$ and the direction of the stress rate l. These tensors and parameters are defined by the following relations:

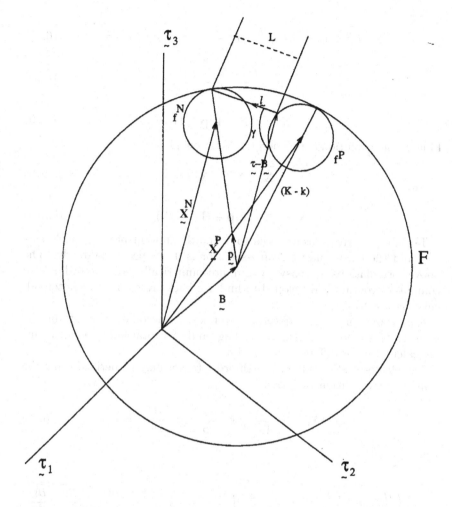

FIGURE 6.2

When yield surface touches the bounding surface.(Voyiadjis and Basuroy-chowdury, 1998)

$$\cos\gamma = \frac{1 : (\tau - \mathbf{B})}{\| 1 : (\tau - \mathbf{B}) \|} \qquad (6.16)$$

$$L = \| (\tau - \mathbf{B}) \| \cos\gamma + \sqrt{(\| \tau - \mathbf{B} \| \cos\gamma)^2 - ((\| \tau - \mathbf{B} \|)^2 - K^2)} \qquad (6.17)$$

$$K^2 = (\| \, \boldsymbol{\tau} - \mathbf{B} \, \|)^2 + L^2 - 2 \, \| \, \boldsymbol{\tau} - \mathbf{B} \, \| \, L \cos\gamma \qquad (6.18)$$

$$\mathbf{p} = \frac{(\boldsymbol{\tau} - \mathbf{B}) + L\mathbf{l}}{K} \qquad (6.19)$$

$$\mathbf{X^N} = (\mathbf{K} - \mathbf{k})\mathbf{p} - \mathbf{B} \qquad (6.20)$$

Finally, the evolution expression for $\mathbf{X}$ is now given as

$$\dot{\mathbf{X}} = \mathbf{X^N} - \mathbf{X^P} \qquad (6.21a)$$

$$\dot{\mathbf{X}} = (K - k)\mathbf{p} - (\mathbf{B} + \mathbf{X^P}) \qquad (6.21b)$$

The above derivations are used to define the movement of the yield surface and the elasto-plastic stiffness tensor is developed accordingly. The above formulations are easy to compute numerically and, accordingly, a computer program is developed to find the backstress, and thereafter calculating the strains.

A computer program is developed to take into account the elasto-plastic stiffness tensor for incremental loading, and the total and plastic strains are calculated for different values of $\boldsymbol{\sigma}$.

The elasto-plastic stiffness fourth-order tensor may be derived using the consistency condition such that

$$\mathbf{D} = \mathbf{E} - \frac{(1 - \beta)}{A} \left(\mathbf{E} : \frac{\partial f}{\partial \boldsymbol{\sigma}} \right) \otimes \left(\frac{\partial f}{\partial \boldsymbol{\sigma}} : \mathbf{E} \right) \qquad (6.22)$$

where

$$A = \left(\frac{\partial f}{\partial \boldsymbol{\sigma}} : \mathbf{E} : \frac{\partial f}{\partial \boldsymbol{\sigma}} \right) (1 - \beta) - \frac{2}{3} C \left(\frac{\partial f}{\partial \boldsymbol{\sigma}} : \frac{\partial f}{\partial \boldsymbol{\sigma}} \right) + \gamma \mathbf{X} : \frac{\partial f}{\partial \boldsymbol{\sigma}} \sqrt{\frac{2}{3} \frac{\partial f}{\partial \boldsymbol{\sigma}} : \frac{\partial f}{\partial \boldsymbol{\sigma}}}$$
$$(6.23)$$

and $\mathbf{E}$ is the fourth order elastic stiffness tensor. In the above expressions, tensor multiplications of fourth-order tensors with second-order tensors using the symbol ":" implies

$$\mathbf{E} : \frac{\partial f}{\partial \boldsymbol{\sigma}} = E_{ijkl} \frac{\partial f}{\partial \sigma_{kl}} \qquad (6.24)$$

Tensor operations on second-order tensors using the symbol $\otimes$ imply

$$\mathbf{a} \otimes \mathbf{b} = \mathbf{a_{ij}b_{kl}} \qquad (6.25)$$

In equation (6.23), the terms C and $\gamma\mathbf{X}$ are given as follows:

$$C = \sum_{i=4}^{4} C^{(i)}, \quad \beta = \sum_{i=1}^{4} \beta^{(i)} \tag{6.26}$$

and

$$\gamma \mathbf{X} = \sum_{i=1}^{4} \gamma^{(i)} \mathbf{X}^{(i)} \tag{6.27}$$

The Lagrangian multiplier $\dot{\Lambda}$ in conjunction with the plastic strain rate tensor given by equation (6.5) is given as follows:

$$\dot{\Lambda} = \frac{1}{A} \left(\frac{\partial f}{\partial \boldsymbol{\sigma}} : \mathbf{E} : \dot{\boldsymbol{\epsilon}} \right) (1 - \beta) \tag{6.28}$$

Isotropic hardening of the type proposed by Chaboche (1989, 1991) in the NLK-T model is used in this chapter. This isotropic hardening rule was initially proposed by Chaboche and generalized by Ohno (1982). The model introduces a surface of nonhardening in the plastic strain space. In the equation for the yield surface, the material constant k is replaced by $k + R$. R is a variable which is zero at the initiation of yielding. The evolution of R is given as follows:

$$\dot{R} = b[Q(q) - R]\dot{p} \tag{6.29}$$

where

$$Q = Q_M + (Q_O + Q_M)e^{-2\mu q} \tag{6.30}$$

Q_O, Q_M, μ and b are material constants. The available q stores one-half the plastic strain amplitude in each cycle $\parallel \Delta\epsilon''/2 \parallel$, which in turn depends on the total strain amplitude to which the material is subjected in cyclic loading.

6.3 Monotonic and Cyclic Tension Loadings on 316 Stainless Steel

The proposed model is first verified by using the available test results obtained by Chaboche (1989, 1991) for monotonic and cyclic tension loadings on 316 stainless steel. Using the cyclic loading history results under stress- and strain-controlled conditions, one can find the metal's deformational properties.

In uniaxial tension-compression loading, the deviatoric stress rate direction always coincides with the normal to the yield surface in this particular example. In addition, since this is a case of proportional loading, the kinematic hardening rule reduces to Phillip's rule (1974), that is $\dot{\tau} = \dot{\mathbf{X}}$. However, it should be noted that for axial proportional loading, the influence of the parameter γ is nullified as shown by equation (6.16), such that

$$l = (\tau - \mathbf{B}) / \| (\tau - \mathbf{B}) \| \tag{6.31}$$

The parameter γ has an effect only in the cases of non-proportional loadings. The values of m and δ given by equations (6.13) and (6.14), respectively, attain a simpler form for the case of axial loading given as follows:

$$m = 2K \tag{6.32}$$

$$\delta = K - (\tau - \mathbf{B}) : l \tag{6.33}$$

The experimental results plotted in Figure 6.3 are obtained by Chaboche (1989, 1991). The value of the coefficients as used by Chaboche (1989, 1991) are given in Table 6.1. For the proposed model, the value of the coefficients is given in Table 2.

TABLE 6.1
Value of the coefficients for the Chaboche model
(1989, 1991). 316 stainless steel.

$C_1 = 80,000$ MPa	$\gamma_1 = 800$	$X = 58$ MPa
$C_2 = 300,000$ MPa	$\gamma_2 = 10,000$	$Q_0 = 14$ MPa
$C_3 = 1,600$ MPa	$\gamma_3 = 0$	$Q_M = 300$ MPa
$C_4 = 17,500$ MPa	$\gamma_4 = 350$	$\mu = 10$
		$b = 8$

The constants C_1, C_2, γ_1 and γ_2 in Table 6.2 are obtained through curve fitting with the uniaxial monotonic experimental curve given by Chaboche (1989, 1991). The strain range in this case is one percent.

The threshold value of X is directly taken from the evaluation of the Chaboche (1989, 1991) model. This constant is determined by Chaboche (1991) from the ratchetting test. He considered the limit to be slightly below the third mean stress level of 60 MPa.

TABLE 6.2

Value of the coefficients for the proposed model. 316 stainless steel.

$C_1 = 80,000$ MPa	$\gamma_1 = 1,000$	$X = 58$ MPa	$\beta_1 = 0.15$
$C_2 = 300,000$ MPa	$\gamma_2 = 10,000$	$Q_0 = 12$ MPa	$\beta_2 = 0.15$
$C_3 = 1,700$ MPa	$\gamma_3 = 0.1$	$Q_M = 600$ MPa	$\beta_3 = 0.15$
$C_4 = 17,000$ MPa	$\gamma_4 = 340$	$\mu = 10$	$\beta_4 = 0.15$
		$b = 8$	

$$E = 187,000 \text{ MPa} \quad \nu = 0.30 \quad k = 122.5 \text{ MPa}$$

The values of C_3, C_4, γ_3 and γ_4 are primarily determined by curve fitting the monotonic uniaxial load Chaboche (1989, 1991) for the later part of the strain range beyond one percent stain. These values of C_3, C_4, γ_3 and γ_4 are further modified to fit with the stress-strain curve for the uniaxial cyclic loading. These final values of the parameters C_3, C_4, γ_3 and γ_4 are distinctly different from those obtained from the uniaxial monotonic tensile curve. Using these final values of the parameters for uniaxial monotonic loadings yield slightly conservative results. However, they predict more accurately the strains for the cyclic loading case.

The coefficients Q_O, Q_M, μ and b in Table 6.2 are based on the constants determined by (Chaboche 1989, 1991) in Table 6.1 for the material of type 316 stainless steel. These constants are further modified in order to predict the uniaxial cyclic strains more accurately. The strain amplitudes are kept identical to those of the experimental results (Figure 6.3). The stress saturation level for the same strain amplitude varies with the different values of Q_O, Q_M, μ and b. These constants are modified so that it best predicts the stress saturation level for a given strain amplitude in which the material is subjected to uniaxial cyclic loading.

The constants β_1, β_2, β_3, β_4 are incorporated in order to predict more accurately the strains. The incorporation of β_1, β_2, β_3, β_4 allows the flexibility of reducing the stress level in the uniaxial monotonic loading and, therefore, provides a better correlation with the experimental observations. The impact of the constants β_1 and β_2 is not as significant as that of β_3 and β_4 on the effect on the predictions of the cyclic loading. The upper bound for β-values is found to be 0.15. If the β-values are taken bigger than 0.15, it may give a better correlation for uniaxial monotonic loading but causes the curved loops in cyclic loading to intersect each other.

For the case of cyclic load, the proposed model shown in Figure 6.4 is compared with the experimental observations shown in Figure 6.3 (1989, 1991). Cyclic loading is performed on the model for different strain rages starting from one percent to 3 percent with an increment of strain rate of 0.5 percent. The loading initiates with a cyclic loading of one percent. The cyclic loading is continued until the saturation of stress occurs. Once this saturation occurs for a particular strain range, the strain range is incremented by 0.5 percent and cyclic loading and unloading is performed until saturation of stress and strain occurs for that particular strain range. This is repeated until cyclic loading is performed for the uniaxial case until a 3 percent strain rage is reached. The proposed model gives a better correlation with the experimental results than that proposed by Chaboche (1989, 1991). The shape of the proposed model in Figure 6.4 for cyclic loading

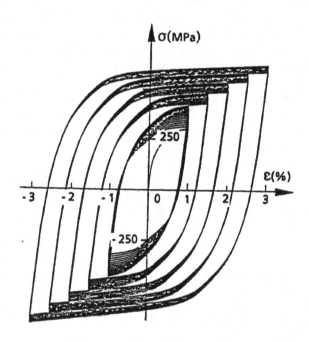

FIGURE 6.3

Experimental results for cyclic stress-strain path with increasing strain levels using the proposed model.(Voyiadjis and Basuroychowdury, 1998)

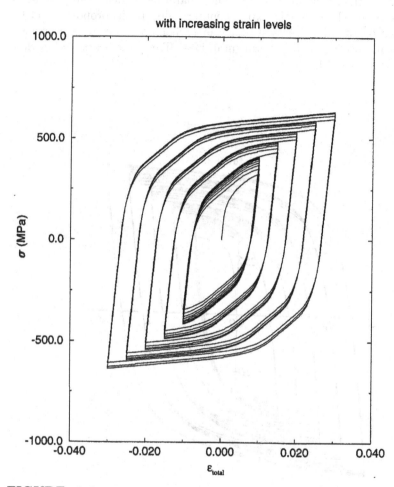

FIGURE 6.4

Simulation of cyclic stress-strain path with increasing strain levels using the proposed model.(Voyiadjis and Basuroychowdury, 1998)

gives a closer correlation with that obtained experimentally in Figure 6.3 than that given by Chaboche (1989, 1991) NLK-T model in Figure 6.5.

The cyclic loading loops using the Chaboche model show a very smooth pattern for loading and unloading in the stress-strain diagram. However, in the proposed model the curves show some tapering at the initiation of yielding which is also reflected by the experimental curves. The proposed model also shows a bigger range of variation of the stresses in a constant strain range than that indicated by the Chaboche model. The proposed model is more in line with the experimental results. In the proposed model, the variation of the stresses start at an earlier point for a given strain range as indicated by the experimental data. This is not captured in the Chaboche model (1991).

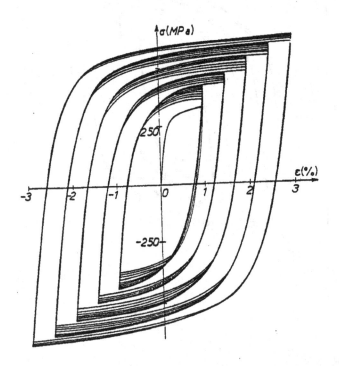

FIGURE 6.5

Simulation of cyclic stress-strain path with increasing strain levels using the Chaboche model (after Chaboche, 1989, 1991; Voyiadjis and Basuroychowdury, 1998)

6.4 Modification of the Kinematic Hardening Model for Non-Proportional Loading

The backstress evolution expressions defined in equations (6.10) are modified here in order to account more appropriately for non-proportional loading paths. The modification involves mainly the replacement of the material constants $\beta^{(i)}$ with functionally dependent parameters $\bar{\beta}^{(i)}$. For non-proportional loading paths, it is clear that the material constant, $\beta^{(i)}$, needs to be continuously adjusted as the stress path changes. The scalar proposed functional expression of $\bar{\beta}^{(i)}$ is expressed in terms of the deviatoric stress tensor direction, the backstress direction, and the direction of the increment of stress. However, for a continuous nonlinear stress path such as a circular stress path, a more involved expression for $\bar{\beta}^{(i)}$ is required. Such an expression needs to account for the dependence not only on the stress path loading direction but also on the rate of change of the stress path direction. The proposed expressions are given in terms of an angle θ expressed in terms of the current increment of stress $\mathbf{l}^C$ and the immediate proceeding increment of stress $\mathbf{l}^P$. The general expressions for $\bar{\beta}^{(i)}$ for continuous nonlinear stress paths are given as follows:

$$\bar{\beta}^{(1)} = (\mathbf{l} : \mathbf{s})(1 - \sin^r \theta)\beta^{(1)} \tag{6.34}$$

$$\bar{\beta}^{(2)} = (\mathbf{l} : \mathbf{s})(1 - \sin^r \theta)\beta^{(2)} \tag{6.35}$$

$$\bar{\beta}^{(3)} = (\mathbf{l} : \mathbf{s}) \sin^r \theta \, \beta^{(3)} \tag{6.36}$$

$$\bar{\beta}^{(4)} = (\mathbf{l} : \mathbf{s}) \sin^r \theta \, \beta^{(4)} \tag{6.37}$$

$$\bar{C}^{(i)} = C^{(i)} \quad i = 1, 2, 4 \tag{6.38}$$

and

$$\bar{C}^{(3)} = C^{(3)} + (A - C^{(3)}) \sin^r \theta \tag{6.39}$$

Very good correlation is obtained with experimental results using the above expressions for continuously non-proportional loading paths such as circular loading paths. The constants A and r in equations (6.34) to (6.39) depend on the type of material used.

The physical interpretation of the scalar $(\mathbf{l} : \mathbf{s})$ is to account for the angle between the stress rate (increment of stress) and the direction of dislocations represented here by $\mathbf{s}$. Initially, the authors used $\mathbf{X}$ instead of $\mathbf{s}$, however, it did not give as good a correlation with the experimental results as, s, did. The proposed expressions presented here account for hardening increases in both cases, when the degree of multiaxiality increases as well as the degree of non-proportionality increases. The degree of nonproportionality is accounted indirectly through the use of the angle, θ.

6.5   Model Predictions and Comparisons with Experimental Data for Non-Proportional Loadings

Stress-strain relations for combined loadings are used here for evaluating the predictions of the model under non-proportional loading conditions. The performance of the model is compared with the experimental data obtained by Shiratori, et al. (1979) conducted on 60/40 brass. These experiments contain extensive non-proportional loading cases with prestraining. Thin-walled tubular specimens of 60/40 brass are first subjected to a prestrain of 4 percent along the axis of the specimen in order to introduce sufficient anisotropy for testing the proficiency of the model. The stress at the end of the 4 percent prestrain is denoted by $\bar{\sigma}_0$. The tube is then subjected to complex loading paths with various stress bends of loading in the axial-torsional stress space.

The axial-torsional subspace may be viewed as a subspace of the Illyushin (1954) five-dimensional deviatoric vector space. The stress vector is defined as follows (Illyushin, 1954):

$$\boldsymbol{\sigma} = \sigma_1 \mathbf{n}_1 + \sigma_3 \mathbf{n}_3 \tag{6.40}$$

where $\mathbf{n}_1$ and $\mathbf{n}_3$ are the orthonormal base vectors in the stress space. The components σ_1 and σ_3 are defined as follows:

$$\sigma_1 = \sigma_{zz} = \sigma \tag{6.41}$$

and

$$\sigma_3 = \sqrt{3}\sigma_{z\theta} = \sqrt{3}\tau \tag{6.42}$$

where z and θ denote, respectively, the tube longitudinal and circumferential directions. The corresponding plastic vector is given as follows:

$$\epsilon'' = \epsilon_1'' \mathbf{n}_1 + \epsilon_3'' \mathbf{n}_3 \tag{6.43}$$

where

$$\epsilon_1'' = \epsilon_{zz}'' \tag{6.44}$$

and

$$\epsilon_3'' = \frac{2}{\sqrt{3}}\epsilon_{z\theta}'' \tag{6.45}$$

The corresponding plastic strain rate vector is defined as follows:

$$\dot{\epsilon}'' = \dot{\epsilon}_1''\mathbf{n}_1 + \dot{\epsilon}_3''\mathbf{n}_3 \tag{6.46}$$

In the axial-torsional subspace, the effective stress is defined such that

$$\sigma_e = (\sigma_1^2 + \sigma_3^2)^{1/2} \tag{6.47}$$

and the corresponding effective plastic strain accumulation is given by

$$\varrho = \int_0^t \sqrt{\frac{2}{3}}\dot{\epsilon}_e''dt \tag{6.48}$$

where

$$\dot{\epsilon}_e'' = (\dot{\epsilon}_1''^2 + \dot{\epsilon}_3''^2)^{1/2} \tag{6.49}$$

The material parameters for 60/40 brass of the proposed model are evaluated in a similar fashion to that outlined earlier for the case of 316 stainless steel. The values of the coefficients are given in Table 6.3.

TABLE 6.3
Value of the coefficients for 60/40 brass.

$C_1 = 5,000$ MPa	$\gamma_1 = 200$	$X = 58$ MPa	$\beta_1 = -0.5$
$C_2 = 135,000$ MPa	$\gamma_2 = 300$	$Q_0 = 12$ MPa	$\beta_2 = -0.5$
$C_3 = 50$ MPa	$\gamma_3 = 0.1$	$Q_M = 600$ MPa	$\beta_3 = 0.125$
$C_4 = 400$ MPa	$\gamma_4 = 5$	$\mu = 10$	$\beta_4 = 0.125$
		$b = 8$	

$$E = 100,000 \text{ MPa} \quad \nu = 0.3 \quad k = 100 \text{ MPa}$$

After the specimens are subjected to a 4 percent prestrain axially, they are then subjected to loading in the σ_3 direction or equivalently to a 90 degree stress bend. Thereafter, σ_e/σ_0 versus ϱ in percent is plotted. Similar experiments are conducted with 120 degree, 135 degree, 150 degree, and 180 degree stress bends after the 4 percent axial prestraining.

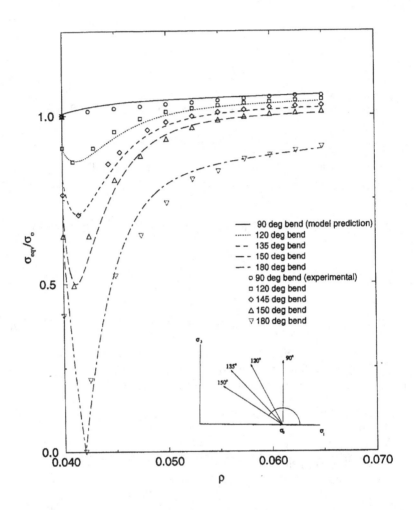

FIGURE 6.6

Model comparison with experimental plastic response with 4% prestrain with 90, 120, 135, 150 and 180 degrees stress bends with $\beta^{(i)}$ given by equations (6.4)-(6.8).(Voyiadjis and Basuroychowdury, 1998)

In Figure 6.6 the response calculated using the proposed model and the experimental observations by Shiratori, et al. (1979) are compared for the different stress bends. The corresponding stress paths are indicated in the figure. In Figure 6.6, the X-axis corresponds to the accumulated plastic strain while the Y-axis corresponds to the equivalent applied stress. Very good correlation is obtained between the response calculated using the proposed model with relations (6.34) - (6.39), and the experimental observations (Shiratori, 1979) for all the stress bends. This is clearly indicated in Figure 6.6.

Figure 6.7 compares the equi-plastic surface as given by the proposed model and the experiments (Shiratori, 1979), respectively. The equi-plastic surfaces are constructed at different plastic strain levels of $\varrho = 0.1, 0.2, 0.5,$ 1.0, 2.0, both for the experimental and the computed model results after a 4 percent prestrain. The plots are made in the stress space.

The proposed model uses an isotropic von Mises yield surface while in reality the yield surface is a distorted anisotropic surface. The use of the von Mises type yield surface avoids the high complexity involved in introducing the anisotropy at the yield surface level. This even becomes more complex when a two-surface anisotropic yield surface model is used and the need is imposed to ensure that the yield surface touches and does not intersect the bounding surface at contact. However, this anisotropy is introduced here through the kinematic hardening with all its resulting complexities. It is clear from the first author's work by Voyiadjis and Foroozesh (1990) and Voyiadjis; et al. (1995) that introducing anisotropy in kinematic hardening is much simpler than introducing it in the yield surface. The only consequence may be a potential inaccuracy in determining the direction of the plastic strain rates at low values of ϱ.

Comparing the experimental results with those obtained using the proposed model in Figure 6.7, one notes the following observations. Referring to Figure 6.7, there is a difference in the shape of the plastic potential surfaces at low values of ϱ which reduces at an increased ϱ since anisotropy is introduced through the kinematic hardening of the model. The proposed additional term in the Armstrong Frederick (1966) model, introduced by the authors, takes care of this additional hardening needed to induce the anisotropic distorted potential surface effect.

6.6 Ratchetting

Ratchetting is the cycle-by-cycle accumulation of the plastic strain for some repetitive loading paths. In Figure 6.8, ratchetting is illustrated using a non-proportional loading history. Tubular specimens of 60/40 brass are used in this case (Shiratori, 1979). The strain accumulation due to non-proportional loading is shown in Figure 6.8 for the proposed plasticity model. The ratchetting path is indicated in Figure 6.8, where $\Delta \epsilon^t = 1\% =$

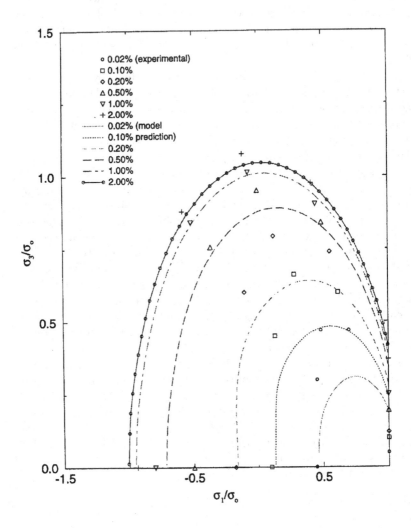

FIGURE 6.7

Equi-plastic surfaces - a comparison.(Voyiadjis and Basuroychowdury, 1998)

$\Delta\gamma/\sqrt{3}$ which is the total strain amplitude normalized to the uniaxial case and $\sigma_d = 67.4\,\text{MPa} = \sigma_1$ which is the constant applied axial stress for the thin walled specimen. The computed response compares reasonably well with the experimental results. However, the decrease in the strain accumulation does not die down as rapidly as in the experimental results.

6.7 Conclusions

A two-surface plasticity model is proposed using a nonlinear hardening rule with evanescent memory evolution. The backstress is calculated as the sum of four components of the type proposed by Chaboche (1989, 1991), the NLK-T model. A new term is added to the Armstrong Frederick (1966) kinematic hardening rule to allow for the motion of the yield surface to be influenced by the direction of the stress rate as proposed by Phillips, et al. (1974). The new term is dependent on the proximity of the yield surface from the bounding surface and on the length of the chord of the bounding surface in the direction of loading.

The proposed model is verified by using the available test results obtained by Chaboche (1989, 1991) for monotonic and cyclic tension loadings on 316 stainless steel. This model gives better correlation with the experimental results than the NLK-T model proposed by Chaboche (1989, 1991).

The material constant $\beta^{(i)}$, used in conjunction with the proposed additional term in the Armstrong Frederick (1966) and Chaboche (1989, 1991) models, is further modified here to account more accurately for nonproportional loading. The constant, $\beta^{(i)}$, is replaced here by a functional scalar expression, $\bar{\beta}^{(i)}$, in terms of the deviatoric stress tensor direction, the backstress direction, and the direction of the increment of stress. This modification allows the variation of the kinematic hardening as the degree of multiaxiality increases. It also accounts for the nonlinear stress path dependence for the material hardening.

The proposed model is tested for non-proportional loading by obtaining numerical results for a series of plastic strain-controlled cyclic tests due to the application of combined axial force and torque to thin-walled tubular specimens of 60/40 brass. The results are compared with the experimental values obtained by Shiratori, et al. (1979). The drift correction due to the finite increments of stress or strain is corrected using an efficient approach that corrects the backstress only. The numerical results obtained compare well with the experimental results (Shiratori, 1979).

The model is also tested for both proportional and non-proportional ratchetting. The computed response compares reasonably well with the experimental results. However, the decrease in the strain accumulation does not die down as rapidly as in the experimental results.

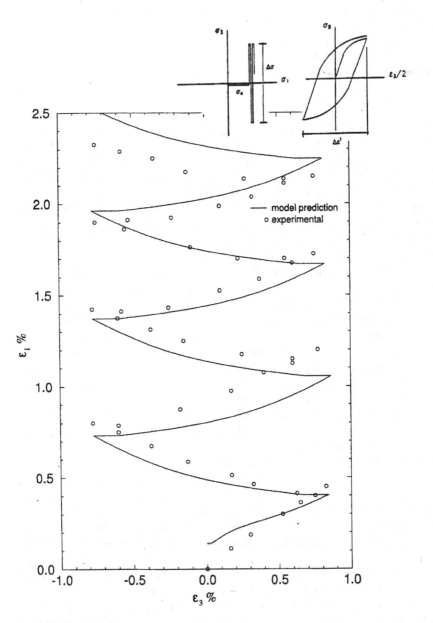

FIGURE 6.8
Model comparison with experimental non-proportional ratchetting.(Voyiadjis and Basuroychowdury, 1998)

7

Coupled Damage Plasticity

A constitutive model is developed in this chapter for anisotropic continuum damage mechanics using finite-strain plasticity. The formulation is given in spatial coordinates (Eulerian reference frame) and incorporates both isotropic and kinematic hardening. The von Mises yield function is modified to include the effects of damage through the use of the hypothesis of elastic energy equivalence. A modified elasto-plastic stiffness tensor that includes the effects of damage is derived within the framework of the proposed model.

It is also shown how the model can be used in conjunction with other damage-related yield criteria. In particular, Gurson's yield function (Gurson 1977) which was later modified by Tvergaard (1982) and Tvergaard and Needleman (1984) is incorporated in the proposed theory. This yield function is derived based on the presence of spherical voids in the material and an evolution law for the void growth is also incorporated. It also shows how a modified Gurson yield function can be related to the proposed model. Some interesting results are obtained in this case. For more details, the reader is referred to Kattan and Voyiadjis (1990, 1993, 2001 a,b), Krajcinovic (1983, 1984), Voyiadjis and Kattan (1996), Voyiadjis and Park (1999), and Wilt and Arnold (1994).

7.1 Stress Transformation between Damaged and Undamaged States

Consider a body in the initial undeformed and undamaged configuration C_o. Let C be the configuration of the body that is both deformed and damaged after a set of external agencies act on it. Next, consider a fictitious configuration of the body $\bar{C}$ obtained from C by removing all the damage that the body has undergone. In other words, $\bar{C}$ is the state of the body after it had only deformed without damage. Therefore, in defining a damage tensor ϕ, its components must vanish in the configuration $\bar{C}$ (see Figure 7.1)

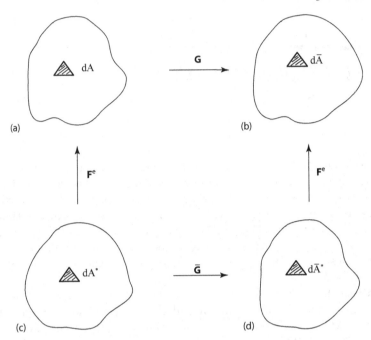

FIGURE 7.1

States of deformation and damage: (a) deformed damage state, (b) fictitious deformed undamaged state, (c) elastically unloaded damaged state (unstressed state), (d) elastically unloaded fictitious undamaged state (fictitious unstressed undamaged state).(Voyiadjis and Kattan, 1999)

7.1.1 Effective Stress Tensor

In the formulation that follows, the Eulerian reference system is used, i.e., all the actual quantities are referred to the configuration C while the effective quantities referred to $\bar{C}$. One first introduces a linear transformation between the Cauchy stress tensor σ and the effective Cauchy stress tensor $\bar{\sigma}$ in the following form:

$$\bar{\sigma}_{ij} = M_{ijkl}\sigma_{kl} \tag{7.1}$$

where M_{ijkl} are the components of the fourth-order linear operator called the damage effect tensor.

A transformation relation for the deviatoric Cauchy stress tensor is next derived. One first writes the deviatoric part σ' in the configuration C as follows:

$$\sigma'_{ij} = \sigma_{ij} - \frac{1}{3}\sigma_{mm}\delta_{ij} \qquad (7.2)$$

where δ_{ij} are the components of the second-order identity tensor **I**. A similar relation exists in the configuration $\bar{C}$ between $\bar{\sigma}$ and $\bar{\sigma}'$ in the following form:

$$\bar{\sigma}'_{ij} = \bar{\sigma}_{ij} - \frac{1}{3}\bar{\sigma}_{mm}\delta_{ij} \qquad (7.3)$$

where δ_{ij} is the same in both C and $\bar{C}$. Substituting for $\bar{\sigma}_{ij}$ from equation (7.1) into equation (7.3) while using equation (7.2), we obtain the following:

$$\bar{\sigma}'_{ij} = M_{ijkl}\sigma'_{kl} + \frac{1}{3}M_{ijmm}\sigma_{nn} - \frac{1}{3}M_{ppqr}\sigma_{qr}\delta_{ij} \qquad (7.4)$$

It is clear from equation (7.4) that a linear relation does not exist between $\bar{\sigma}'$ and σ'. On the other hand, one might suspect that the last two terms on the right-hand side of the equation (7.4) cancel each other when they are written in expanded form. However, this possibility can be easily dismissed as follows: suppose one assumes $\bar{\sigma}'_{ij} = M_{ijkl}\sigma'_{kl}$. Using this with equation (7.4) one concludes that $M_{ijkl}\sigma_{nn} = M_{ppqr}\sigma_{qr}\delta_{ij}$. Now consider the case when $i \neq j$. One has $\delta_{ij} = 0$ and, therefore, $M_{ijmm}\sigma_{nn} = 0$. It is clear that this is a contradiction to the fact that generally $M_{ijmm} \neq 0$ and $\sigma_{nn} \neq 0$. Therefore, the additional terms in the equation (7.4) are non-trivial and such a linear transformation cannot be assumed.

Upon examining equation (7.4) in more detail, eliminating σ'_{kl} by using equation (7.2) and simplifying the resulting expression, one obtains the following:

$$\bar{\sigma}'_{ij} = N_{ijkl}\sigma_{kl} \qquad (7.5)$$

where N_{ijkl} are the components of a fourth-order tensor given by:

$$N_{ijkl} = M_{ijkl} - \frac{1}{3}M_{rrkl}\delta_{ij} \qquad (7.6)$$

Equation (7.5) represents a linear transformation between the effective deviatoric Cauchy stress tensor $\bar{\sigma}'$ and the Cauchy stress tensor σ. However, in this case the linear operator **N** is not simply the damage effect tensor **M** but a linear function of **M** as shown in equation (7.6). The tensors **M** and **N** are mappings $S \rightarrow \bar{S}$ and $S \rightarrow \bar{S}_{dev.}$, respectively, where S is the stress space in the current configuration C and $\bar{S}$ is the stress space in the fictitious undamaged state $\bar{C}$, with $\sigma \, \epsilon \, S$ and $\bar{\sigma} \, \epsilon \, \bar{S}$.

Next, we consider the effective stress invariants and their transformation in the configuration $\bar{C}$. It is seen from equation (7.5) that the first effective deviatoric stress invariant $\bar{\sigma}'_{ii}$ is given by:

$$\bar{\sigma}'_{ii} = N_{iikl}\sigma_{kl} = 0 \qquad (7.7)$$

since $N_{iikl} = 0$ by direct contraction in equation (7.6). Therefore, one obtains $\bar{\sigma}'_{ii} = \sigma'_{ii} = 0$.

The problem becomes more involved when considering the effective stress invariant $\bar{\sigma}'_{ij}\bar{\sigma}'_{ij}$. Using equation (7.4) along with equation (7.2), one obtains:

$$\bar{\sigma}'_{ij}\bar{\sigma}'_{ij} = A_{klmn}\sigma'_{kl}\sigma'_{mn} + B_{pq}\sigma'_{pq} + C \qquad (7.8)$$

where

$$A_{klmn} = M_{ijkl}M_{ijmn} - \frac{1}{3}M_{ttmn}M_{rrkl} \qquad (7.9a)$$

$$B_{pq} = \frac{2}{3}\sigma_{mm}(M_{ijkk}M_{ijpq} - \frac{1}{3}M_{ttpq}M_{rrnn}) \qquad (7.9b)$$

$$C = \frac{1}{2}\sigma^2_{mm}(M_{ijpp}M_{ijqq} - \frac{1}{3}M_{ttmm}M_{rrnn}) \qquad (7.9c)$$

Substituting for $\boldsymbol{\sigma}'$ from equation (7.2) into equation (7.8) [or more directly using equation (7.5) along with equation (7.6)], one obtains:

$$\bar{\sigma}'_{ij}\bar{\sigma}'_{ij} = H_{klmn}\sigma_{kl}\sigma_{mn} \qquad (7.10)$$

where the fourth-order tensor $\mathbf{H}$ is given by:

$$H_{klmn} = N_{ijkl}N_{ijmn} \qquad (7.11)$$

and the tensor $\mathbf{N}$ is given by equation (7.6). The transformation equation (7.10) will be used in the next sections to transform the von Mises yield criterion into the configuration $\bar{C}$.

7.1.2   Effective Backstress Tensor

In the theory of plasticity, kinematic hardening is modeled by the motion of the yield surface in the stress space. This is implemented mathematically by the evolution of the shift or backstress tensor $\mathbf{X}$. The backstress tensor $\mathbf{X}$ denotes the position of the center of the yield surface in the stress space. For this purpose, one studies now the transformation of this tensor in the configuration C and $\bar{C}$.

Let $\mathbf{X}'$ be the deviatoric part of $\mathbf{X}$. Therefore, one has the following relation:

$$X'_{ij} = X_{ij} - \frac{1}{3}X_{mm}\delta_{ij} \qquad (7.12)$$

where both $\mathbf{X}$ and $\mathbf{X}'$ are referred to the configuration C. Let their effective counterparts $\bar{\mathbf{X}}$ and $\bar{\mathbf{X}}'$ be referred to the configuration $\bar{C}$. Similarly to equation (7.12), we have:

$$\bar{X}'_{ij} = \bar{X}_{ij} - \frac{1}{3}\bar{X}_{mm}\delta_{ij} \tag{7.13}$$

Assuming a linear transformation (based on the same argument used for the stresses) similar to equation (7.1) between the effective backstress tensor $\bar{\mathbf{X}}$ and the backstress tensor $\mathbf{X}$:

$$\bar{X}_{ij} = M_{ijkl}X_{kl} \tag{7.14}$$

and following the same procedure in the derivation of equation (7.5), we obtain the following linear transformation between $\mathbf{X}$ and $\bar{\mathbf{X}}'$:

$$\bar{X}'_{ij} = N_{ijkl}X_{kl} \tag{7.15}$$

The effective backstress invariants have similar forms to those of the effective stress invariants, mainly, $\bar{X}'_{ii} = X'_{ii} = 0$, and

$$\bar{X}'_{ij}\bar{X}'_{ij} = H_{klmn}X_{kl}X_{mn} \tag{7.16}$$

In addition, one more transformation equation needs to be given before we can proceed to the constitutive model. By following the same procedure for the other invariants, the mixed invariant $\sigma_{ij}X_{ij}$ in the configuration C is transformed to $\bar{\sigma}'_{ij}\bar{X}'_{ij}$ as follows:

$$\bar{\sigma}'_{ij}\bar{X}'_{ij} = H_{klmn}\sigma_{kl}X_{mn} \tag{7.17}$$

and a similar relation holds for the invariant $\bar{X}'_{ij}\bar{\sigma}'_{ij}$. The stress and backstress transformation equations will be used later in the constitutive model.

7.2 Strain Rate Transformation between Damaged and Undamaged States

In the general elasto-plastic analysis of deforming bodies, the spatial strain rate tensor $\mathbf{d}$ in the configuration C is decomposed additively as follows (Nemat-Nasser, 1979, 1983; Lee, 1981):

$$d_{ke} = d'_{kl} + d''_{kl} \tag{7.18}$$

where $\mathbf{d}'$ and $\mathbf{d}''$ denote the elastic and plastic parts of $\mathbf{d}$, respectively. In equation (7.18), the assumption of small elastic strains is made, however, finite plastic deformations are allowed. On the other hand, the decomposition in equation (7.18) will be be true for any amount of elastic strain if the physics of elasto-plasticity is invoked, for example the case of single crystals. A thorough account of this is given by Asaro (1983).

In the next two subsections, the necessary transformation equations between the configurations C and $\bar{C}$ will be derived for the elastic strain and plastic strain rate tensors. In this derivation, it is assumed that the elastic strains are small compared with the plastic strains and, consequently, the elastic strain tensor is taken to be the usual engineering elastic strain tensor ϵ'. In addition, it is assumed that an elastic strain energy function exists such that a linear relation can be used between the Cauchy stress tensor σ and the engineering elastic strain tensor ϵ'. The tensor ϵ' is defined here as the linear term of the elastic part of the spatial strain tensor where second-order terms are neglected. For more details, see the work by Kattan and Voyiadjis (1992) and Voyiadjis and Kattan (1999).

7.2.1   Effective Elastic Strain

The elastic constitutive equation to be used is based on one of the assumptions outlined in the previous paragraph and is represented by the following linear relation in the configuration $\bar{C}$:

$$\bar{\sigma}_{ij} = \bar{E}_{ijkl}\bar{\epsilon}'_{kl} \tag{7.19}$$

where $\bar{\mathbf{E}}$ is the fourth-rank elasticity tensor given by:

$$\bar{E}_{ijkl} = \lambda\delta_{ij}\delta_{kl} + G(\delta_{ik}\delta_{jl} + \delta_{il}\delta_{jk}) \tag{7.20}$$

where λ and M are Lame's constants. Based on the constitutive equation (7.19), the elastic strain energy function $U(\epsilon', \phi)$ in the configuration $\bar{C}$ is given by:

$$U(\bar{\epsilon}', \mathbf{0}) = \frac{1}{2}\bar{E}_{ijkl}\bar{\epsilon}'_{ij}\bar{\epsilon}'_{kl} \tag{7.21}$$

One can now define the complementary elastic energy function $V(\sigma, \phi)$, based on a Legendre transformation, as follows:

$$V(\sigma, \phi) = \sigma_{ij}\epsilon'_{ij} - U(\bar{\epsilon}', \phi) \tag{7.22}$$

By taking the partial derivative of equation (7.22) with respect to the stress tensor σ, one obtains:

$$\epsilon'_{ij} = \frac{\partial V(\sigma, \phi)}{\partial \sigma_{ij}} \tag{7.23}$$

Substituting expression (7.21) into equation (7.22) in the configuration $\bar{C}$, one obtains the following expression for $V(\sigma, \phi)$ in the configuration $\bar{C}$ as follows:

$$V(\bar{\sigma}, 0) = \frac{1}{2}\bar{E}_{ijkl}^{-1}\bar{\sigma}_{ij}\bar{\sigma}_{kl} \qquad (7.24)$$

The hypothesis of elastic energy equivalence, which was initially proposed by Sidoroff (1981), is now used to obtain the required relation between ϵ' and $\bar{\epsilon}'$. In this hypothesis, one assumes that the elastic energy $V(\sigma, \phi)$ in the configuration C is equivalent in form to $V(\bar{\sigma}, 0)$ in the configuration $\bar{C}$. Therefore, one writes:

$$V(\sigma, \phi) = V(\bar{\sigma}, 0) \qquad (7.25)$$

where $V(\sigma, \phi)$ is the complementary elastic energy in C and is given by:

$$V(\sigma, \phi) = \frac{1}{2}E_{ijkl}^{-1}(\phi)\sigma_{ij}\sigma_{kl} \qquad (7.26)$$

where the superscript -1 indicates the inverse of the tensor.

In the equation (7.26), the effective elasticity modulus $\mathbf{E}(\phi)$ is a function of the damage tensor ϕ and is no longer a constant. Using equation (7.25) along with expressions (7.24) and (7.26), one obtains the following relation between $\bar{\mathbf{E}}$ and $\mathbf{E}(\phi)$:

$$E_{klmn}(\phi) = M_{ijkl}^{-1}(\phi)\bar{E}_{ijpq}M_{pqmn}^{-T}(\phi) \qquad (7.27)$$

where the superscript $-T$ indicates the transpose of the inverse of the tensor as depicted by $M_{ijkl}^{-1}M_{mnkl} = \delta_{im}\delta_{jn}$. Finally, using equation (7.23) along with equations (7.24), (7.25), and (7.27), one obtains the desired linear relation between the elastic strain tensor ϵ' and its effective counterpart $\bar{\epsilon}'$:

$$\bar{\epsilon}_{kl}' = M_{klmn}^{-T}\epsilon_{mn}' \qquad (7.28)$$

The two transformation equations (7.27) and (7.28) will be incorporated later in this chapter in the general inelastic constitutive model that will be developed.

7.2.2 Effective Plastic Strain Rate

The constitutive model to be developed here is based on a von Mises type yield function $f(\sigma', \mathbf{X}', \kappa, \phi)$ in the configuration C that involves both isotropic and kinematic hardening through the evolution of the plastic work κ and the backstress tensor $\mathbf{X}'$, respectively. The corresponding yield function $f(\bar{\sigma}', \bar{\mathbf{X}}', \bar{\kappa}, 0)$ in the configuration $\bar{C}$ is given by:

$$f = \frac{3}{2}(\bar{\sigma}'_{kl} - \bar{X}'_{kl})(\bar{\sigma}'_{kl} - \bar{X}'_{kl}) - \bar{\sigma}_o^2 - c\bar{\kappa} = 0 \qquad (7.29)$$

where σ_o and c are material parameters denoting the uniaxial yield strength and isotropic hardening, respectively. The plastic work κ is a scalar function and its evolution in the configuration $\bar{C}$, is taken here to be in the following form:

$$d\bar{\kappa} = \sqrt{\bar{d}''_{ij}\bar{d}''_{ij}} \qquad (7.30)$$

where d''_{ij} is the plastic part of the spatial strain rate tensor $\mathbf{d}$.

Isotropic hardening is described by the evolution of the plastic work κ as given above. In order to describe kinematic hardening, the Prager-Ziegler evolution law (Oldroyd, 1950) is used here in the configuration $\bar{C}$, as follows:

$$d\bar{\alpha}_{ij} = d\bar{\mu}(\bar{\sigma}'_{ij} - \bar{X}'_{ij}) \qquad (7.31)$$

where $d\bar{\mu}$ is a scalar function to be determined shortly.

The plastic flow in the configuration $\bar{C}$ is described by the associated flow rule in the form:

$$\bar{d}''_{ij} = d\bar{\Lambda}\frac{\partial f}{\partial \bar{\sigma}_{ij}} \qquad (7.32)$$

where $d\bar{\Lambda}$ is a scalar function introduced as a Lagrange multiplier in the constraint thermodynamic equations (see section 7.3), that is still to be determined. In the present formulation, it is assumed that the associated flow of plasticity will still hold in the configuration C, that is:

$$d''_{ij} = d\Lambda\frac{\partial f}{\partial \sigma_{ij}} \qquad (7.33)$$

where $d\Lambda$ is another scalar function that is to be determined.

Substituting the yield function f of equation (7.29) into equation (7.32) and using the transformation equations (7.5) and (7.15), one obtains:

$$\bar{d}''_{ij} = 3d\bar{\Lambda}N_{ijkl}(\sigma_{kl} - X_{kl}) \qquad (7.34)$$

On the other hand, substituting the yield function f of equation (7.29) into equation (7.33) and noting the appropriate transformation (7.10) and (7.16), one obtains:

$$d''_{ij} = 3d\Lambda H_{ijkl}(\sigma_{kl} - X_{kl}) \qquad (7.35)$$

It is noticed that plastic incompressibility exists in the configuration $\bar{C}$ as seen from equation (7.34) where $\bar{d}''_{mm} = 0$ since $N_{mmkl} = 0$. However, this is not true in the configuration C since d''_{mm} does not vanish depending on H_{mmkl} as shown in equation (7.35).

In order to derive the transformation equation between $\mathbf{d}''$ and $\bar{\mathbf{d}}''$, one first notices that:

$$\frac{\partial f}{\partial \sigma_{ij}} = \frac{\partial f}{\partial \bar{\sigma}_{pq}} \frac{\partial \bar{\sigma}_{pq}}{\partial \sigma_{ij}} = \frac{\partial f}{\partial \bar{\sigma}_{pr}} M_{pqij} \tag{7.36}$$

where M_{pqij} is defined in the equation (7.1) as $\partial \bar{\sigma}_{pq}/\partial \sigma_{ij}$. Using the above relation along with equations (7.32) and (7.33), one obtains:

$$\bar{d}''_{ij} = \frac{d\bar{\Lambda}}{d\Lambda} M^{-1}_{ijmn} d''_{mn} \tag{7.37}$$

The above equation represents the desired relation, except that the expression $d\bar{\Lambda}/d\Lambda$ needs to be determined. This is done by finding explicit expressions for both $d\Lambda$ and $d\bar{\Lambda}$ using the consisting conditions. The rest of this section is devoted to this task. But first one needs to determine an appropriate expression for $d\bar{\mu}$ that appears in equation (7.31), since it plays an essential role in the determination of $d\bar{\Lambda}$.

In order to determine an expression for $d\bar{\mu}$, one assumes that the projection of $d\bar{\mathbf{X}}'$ on the gradient of the yield surface f in the stress space is equal to $b\bar{\mathbf{d}}''$ in the configuration $\bar{C}$, where b is a material parameter to be determined from the uniaxial tension test (Voyiadjis, 1984; Voyiadjis and Kiousis, 1987). This assumption is written as follows:

$$b\bar{d}''_{kl} = d\bar{X}'_{mn} \frac{\frac{\partial f}{\partial \bar{\sigma}_{mn}}}{\frac{\partial f}{\partial \bar{\sigma}_{pq}} \frac{\partial f}{\partial \bar{\sigma}_{pq}}} \frac{\partial f}{\partial \bar{\sigma}_{kl}} \tag{7.38}$$

Substituting for $d\bar{\mathbf{X}}'$ and $\mathbf{d}''$ from equation (7.31) and (7.32), respectively, into equation (7.38) and post-multiplying the resulting equation by $\partial f/\partial \bar{\sigma}_{kl}$, one obtains the required expression for $d\bar{\mu}$:

$$d\bar{\mu} = bd\bar{\Lambda} \frac{\frac{\partial f}{\partial \bar{\sigma}_{mn}} \frac{\partial f}{\partial \bar{\sigma}_{mn}}}{(\bar{\sigma}'_{pq} - \bar{X}'_{pq}) \frac{\partial f}{\partial \bar{\sigma}_{pq}}} \tag{7.39}$$

Using the elastic linear relationship in equation (7.19), and taking its rate, one obtains:

$$d\bar{\sigma}_{ij} = \bar{E}_{ijkl} \bar{d}'_{kl} \tag{7.40}$$

where $\bar{d}'_{kl}$ is assumed to be equal to $d\bar{\epsilon}'_{kl}$ based on the assumption of small elastic strains as discussed in section 7.2.1. Eliminating $\bar{\mathbf{d}}'$ from equation (7.40) through the use of expressions (7.18) and (7.32), one obtains:

$$d\bar{\sigma}_{ij} = \bar{E}_{ijkl}\left(\bar{d}_{kl} - d\bar{\Lambda}\frac{\partial f}{\partial \bar{\sigma}_{kl}}\right) \tag{7.41}$$

The scalar multiplier $d\bar{\Lambda}$ is obtained from the consisting condition $df(\bar{\sigma}'_{kl}, \bar{X}'_{kl}, 0, \bar{\kappa}) = 0$ such that:

$$\frac{\partial f}{\partial \bar{\sigma}'_{kl}}d\bar{\sigma}'_{kl} + \frac{\partial f}{\partial \bar{X}'_{kl}}d\bar{X}'_{kl} + \frac{\partial f}{\partial \bar{\kappa}}d\bar{\kappa} = 0 \tag{7.42}$$

Using equations (7.18), (7.30), (7.32), (7.39), and (7.41) into equation (7.42), one obtains the following expression for $d\bar{\Lambda}$:

$$d\bar{\Lambda} = \frac{1}{\bar{Q}}\frac{\partial f}{\partial \bar{\sigma}'_{kl}}\bar{E}_{klmn}\bar{d}_{mn} \tag{7.43}$$

where $\bar{Q}$ is scalar and given by:

$$\bar{Q} = \frac{\partial f}{\partial \bar{\sigma}'_{ab}}\bar{E}_{abcd}\frac{\partial f}{\partial \bar{\sigma}'_{cd}} - \frac{\partial f}{\partial \bar{\kappa}}\sqrt{\frac{\partial f}{\partial \bar{\sigma}_{pq}}\frac{\partial f}{\partial \bar{\sigma}_{pq}}} - b\frac{\partial f}{\partial \bar{X}'_{ef}}(\bar{\sigma}'_{ef} - \bar{X}'_{ef})\frac{\frac{\partial f}{\partial \bar{\sigma}_{ij}}\frac{\partial f}{\partial \bar{\sigma}_{ij}}}{(\bar{\sigma}'_{gh} - \bar{X}'_{gh})\frac{\partial f}{\partial \bar{\sigma}_{gh}}} \tag{7.44}$$

Assuming that the Prager-Ziegler kinematic hardening rule holds in the configuration C along with the projection assumption of equation (7.13), one can derive a similar equation to (7.43), in the following form:

$$d\Lambda = \frac{1}{Q}\frac{\partial f}{\partial \sigma'_{kl}}E_{klmn}d_{mn} \tag{7.45}$$

where Q is given by:

$$Q = \frac{\partial f}{\partial \sigma'_{ab}}E_{abcd}\frac{\partial f}{\partial \sigma'_{cd}} - \frac{\partial f}{\partial \kappa}\sqrt{\frac{\partial f}{\partial \sigma_{pq}}\frac{\partial f}{\partial \sigma_{pq}}} - b\frac{\partial f}{\partial X'_{ef}}(\sigma'_{ef} - X'_{ef})\frac{\frac{\partial f}{\partial \sigma_{ij}}\frac{\partial f}{\partial \sigma_{ij}}}{(\sigma'_{gh} - X'_{gh})\frac{\partial f}{\partial \sigma_{gh}}} \tag{7.46}$$

In contrast to the method used by Voyiadjis and Kattan (1990) where the two yield functions in the configurations C and $\bar{C}$ are assumed to be equal, a more consistent approach is adopted here. This approach is based on the assumptions used to derive equation (7.45). It is clear that in this method, the two yield functions in the configurations C and $\bar{C}$ are treated separately and two separate consistency conditions are thus invoked. In the authors' opinion, this emphasizes a more consistent approach than the method used by Voyiadjis and Kattan (1990).

One is now left with the tedious algebraic manipulations of equations (7.43) and (7.45) in order to derive an appropriate form for the ratio $d\bar{\Lambda}/d\Lambda$.

First, equation (7.45) is rewritten in the following form, where the appropriate transformations $\bar{\mathbf{E}} \rightarrow \mathbf{E}$ and $\boldsymbol{\sigma} \rightarrow \bar{\boldsymbol{\sigma}}$ are used:

$$Q d\Lambda = \frac{\partial f}{\partial \bar{\sigma}'_{ij}} \bar{E}_{ijpq} M^{-T}_{pqmn} d_{mn} \tag{7.47}$$

Then, one expands equation (7.43) by using the appropriate transformations $\bar{\mathbf{d}}' \rightarrow \mathbf{d}'$ and $\bar{\mathbf{d}}'' \rightarrow \mathbf{d}''$ to obtain:

$$d\bar{\Lambda} = \frac{1}{\bar{Q}} \frac{\partial f}{\partial \bar{\sigma}'_{kl}} \bar{E}_{klmn} [dM^{-T}_{mnpq} E^{-1}_{ijpq} \sigma_{ij} + M^{-T}_{mnpq} d_{pq} - (M^{-T}_{mnpq} - \frac{d\bar{\Lambda}}{d\Lambda} M^{-1}_{mnpq}) d''_{pq}] \tag{7.48}$$

where the derivative dM_{mnpq} is defined in the next section. The last major step in the derivation is to substitute the term on the right-hand side of equation (7.47) for the results using the transformations $\boldsymbol{\sigma} \rightarrow \bar{\boldsymbol{\sigma}}, \boldsymbol{\sigma} \rightarrow \bar{\boldsymbol{\sigma}}', \mathbf{E} \rightarrow \bar{\mathbf{E}}$, and others. Once this is done, the following relation is obtained:

$$(\bar{Q} - \frac{\partial f}{\partial \bar{\sigma}'_{kl}} \bar{E}_{klmn} \frac{\partial f}{\partial \bar{\sigma}_{mn}}) d\bar{\Lambda} = (Q - \frac{\partial f}{\partial \sigma'_{ij}} E_{ijpq} \frac{\partial f}{\partial \sigma_{pq}}) d\Lambda$$
$$+ \frac{\partial f}{\partial \bar{\sigma}'_{kl}} \bar{E}_{klmn} dM^{-T}_{mnpq} E^{-1}_{ijpq} \sigma_{ij} \tag{7.49}$$

The above equation is rewritten in the form:

$$a_1 d\bar{\Lambda} = a_2 d\Lambda + a_3 \tag{7.50}$$

where

$$a_1 = \bar{Q} - \frac{\partial f}{\partial \bar{\sigma}'_{kl}} \bar{E}_{klmn} \frac{\partial f}{\partial \bar{\sigma}_{mn}} \tag{7.51a}$$

$$a_2 = Q - \frac{\partial f}{\partial \sigma'_{kl}} E_{klmn} \frac{\partial f}{\partial \sigma_{mn}} \tag{7.51b}$$

$$a_3 = \frac{\partial f}{\partial \bar{\sigma}'_{kl}} \bar{E}_{klmn} dM^{-T}_{mnpq} E^{-1}_{ijpq} \sigma_{ij} \tag{7.51c}$$

It is noticed that a_1 and a_2 are the last two terms on the right-hand sides of equations (7.44) and (7.46), respectively.

It should be noted that when the material undergoes only plastic deformation without damage, that is when the configurations C and $\bar{C}$ coincide, then $a_1 = a_2$ and $a_3 = 0$ since $d\mathbf{M}$ vanishes in this case, thus leading to $d\bar{\Lambda} = d\Lambda$.

The relation (7.50) is now substituted into equation (7.37) along with equation (7.35) to obtain the following nonlinear transformation equation for the plastic part of the spatial strain rate tensor:

$$\bar{d}''_{ij} = X_{ijkl} d''_{kl} + Z_{ij} \tag{7.52}$$

where the tensors $\mathbf{X}$ and $\mathbf{Z}$ are given by:

$$X_{ijkl} = \frac{a_2}{a_1} M^{-1}_{ijkl} \tag{7.53}$$

$$Z_{ij} = 3\frac{a_3}{a_1} M^{-1}_{ijkl} H_{klmn}(\sigma_{mn} - X_{mn}) \tag{7.54}$$

The transformation equation (7.52) will be used later in the derivation of the constitutive equations.

7.3   Constitutive Model

In this section, a coupled constitutive model will be derived incorporating both elasto-plasticity and damage. This section is divided into three subsections detailing the derivation starting with the equations then proceeding to the desired coupling.

7.3.1   Damage Evolution

In this section, an inelastic constitutive model is derived in conjunction with the damage transformation equations presented in the previous sections. An elasto-plasticity stiffness tensor that involves damage effects is derived in the Eulerian reference system. In this formulation, rate-dependent effects are neglected and isothermal conditions are assumed. The damage evolution criterion to be used here is that proposed by Lee et al. (1985) and is given by:

$$g(\bar{\sigma}, L) = \frac{1}{2} J_{ijkl} \bar{\sigma}_{ij} \bar{\sigma}_{kl} - l_o^2 - L(l) \equiv 0 \tag{7.55}$$

where J_{ijkl} are the components of a constant fourth-order tensor that is symmetric and isotopic. This tensor is represented by the following matrix:

$$
\mathbf{J} \equiv \begin{bmatrix}
1 & \mu & \mu & 0 & 0 & 0 \\
\mu & 1 & \mu & 0 & 0 & 0 \\
\mu & \mu & 1 & 0 & 0 & 0 \\
0 & 0 & 0 & 2(1-\mu) & 0 & 0 \\
0 & 0 & 0 & 0 & 2(1-\mu) & 0 \\
0 & 0 & 0 & 0 & 0 & 2(1-\mu)
\end{bmatrix} \tag{7.56}
$$

where μ is a material constant satisfying $-\frac{1}{2} \le \mu \le 1$. In equation (7.55), l_o represents the initial damage threshold, $L(l)$ is the increment of damage threshold, and l is the scalar variable that represents overall damage.

During the process of plastic deformation and damage, the power of dissipation Π is given by (Voyiadjis and Kattan, 1990):

$$
\Pi = \sigma_{ij} d_{ij}'' + \sigma_{kl} d\phi_{kl} - L dl \tag{7.57}
$$

In order to obtain the actual values of the parameters $\boldsymbol{\sigma}, \boldsymbol{\phi}, \kappa$, and l, one needs to solve an extremization problem, i.e, the power of dissipation Π is to be extremized subject to two constraints, namely, $f(\boldsymbol{\sigma}', \mathbf{X}', \kappa, \boldsymbol{\phi}) = 0$ and $g(\bar{\boldsymbol{\sigma}}, L) = 0$. Using the method of the calculus of functions of several variables, one introduces two Lagrange multipliers $d\lambda_1$, and $d\lambda_2$, and forms the function Ψ such that:

$$
\Psi = \Pi - d\lambda_1 f - d\lambda_2 g \tag{7.58}
$$

The problem now reduces to that of extremizing the function Ψ. Therefore, one uses the necessary conditions $\partial \Psi / \partial \boldsymbol{\sigma} = 0$ and $\partial \Psi / \partial L = 0$ to obtain:

$$
d_{ij}'' + d\phi_{ij} - d\lambda_1 \frac{\partial f}{\partial \sigma_{ij}} - d\lambda_2 \frac{\partial g}{\partial \sigma_{ij}} = 0 \tag{7.59a}
$$

$$
-dl - d\lambda_2 \frac{\partial g}{\partial L} = 0 \tag{7.59b}
$$

Next, one obtains from equation (7.55) that $\partial g / \partial L = -1$. Substituting this into equation (7.59b), one obtains $d\lambda_2 = dl$. Thus $d\lambda_2$ describes the evolution of the overall damage parameter l which is to be determined shortly. Using equation (7.59b) and assuming that damage and plastic deformation are two independent processes, one obtains the following two rate equations for the plastic strain and damage tensors:

$$
d_{ij}'' = d\lambda_1 \frac{\partial f}{\partial \sigma_{ij}} \tag{7.60a}
$$

$$d\phi_{ij} = dl\frac{\partial g}{\partial \sigma_{ij}} \tag{7.60b}$$

The first equation of (7.60) is the associated flow rule for the plastic strain introduced earlier in equation (7.33), while the second is the evolution of the damage tensor. It is to be noted that $d\lambda_1$ is exactly the same as the multiplier $d\Lambda$ used earlier. However, one needs to obtain explicit expressions for the multipliers $d\Lambda$ and dl. The derivation of an expression for $d\Lambda$ will be left for the next section when the inelastic constitutive model is discussed. Now one proceeds to derive an expression for dl. This is done by invoking the consistency condition $dg(\boldsymbol{\sigma}, \boldsymbol{\phi}, L) = 0$. Therefore, one obtains:

$$\frac{\partial g}{\partial \sigma_{ij}}d\sigma_{ij} + \frac{\partial g}{\partial \phi_{kl}}d\phi_{kl} + \frac{\partial g}{\partial L}dL = 0 \tag{7.61}$$

Substituting for $d\phi$ from equation (7.60b) along with $\partial g/\partial L = -1$ and $dL = dl(\partial L/\partial l)$, we obtain:

$$dl = \frac{\frac{\partial g}{\partial \sigma_{ij}}d\sigma_{ij}}{\frac{\partial L}{\partial l} - \frac{\partial g}{\partial \phi_{pq}}\frac{\partial g}{\partial \sigma_{pq}}} \tag{7.62}$$

Finally, by substituting equation (7.62) into equation (7.60b), we obtain the general evolution equation for the damage tensor $\boldsymbol{\phi}$ as follows:

$$d\phi_{kl} = \frac{\frac{\partial g}{\partial \sigma_{ij}}d\sigma_{ij}}{\frac{\partial L}{\partial l} - \frac{\partial g}{\partial \phi_{pq}}\frac{\partial g}{\partial \sigma_{pq}}}\frac{\partial g}{\partial \sigma_{kl}} \tag{7.63}$$

The evolution equation (7.63) is to be incorporated in the constitutive model in the next two sections. It will also be used in the derivation of the elasto-plastic stiffness tensor. It should be noted that equation (7.63) is based on the damage criterion of equation (7.55) which is applicable to anisotropic damage. However, using the form for $\mathbf{J}$ given in equation (7.56) restricts the formulation to isotropy.

7.3.2 Plastic Deformation

In the analysis of finite strain plasticity, one needs to define an appropriate corotational stress rate that is objective and frame-indifferent. Detailed discussions of the types of stress rates are available in the papers of Voyiadjis and Kattan (1989) and Paulun and Pecherski (1985). The corotational stress rate to be adopted in this model is given for $\boldsymbol{\sigma}$ in the following form:

$$\overset{\circ}{\sigma}_{ij} = d\sigma_{ij} - \Omega_{ip}\sigma_{pj} + \sigma_{iq}\Omega_{qj} \tag{7.64}$$

where the modified spin tensor $\mathbf{\Omega}$ is given by (Paulun and Pecherski, 1985; Voyiadjis and Kattan, 1989) as follows:

$$\Omega_{ij} = \omega W_{ij} \tag{7.65}$$

In equation (7.65), $\mathbf{W}$ is the material spin tensor (the antisymmetric part of the velocity gradient) and ω is an influence scalar function too be determined. The effect of ω on the evolution of the stress and backstress is discussed in detail by Voyiadjis and Kattan (1989). The corotational rate $\overset{\circ}{\boldsymbol{\alpha}}$ has a similar expression as that in equation (7.64) keeping in mind that the modified spin tensor $\mathbf{\Omega}$ remains the same in both equations.

The yield function to be used in this model is the function f given by equation (7.29) with both isotropic and kinematic hardening. Isotropic hardening is described by the evolution of the plastic work as given earlier by equation (7.30), while kinematic hardening is given by equation (7.31). Most of the necessary plasticity equations were given in section 7.2.2 and the only thing remaining is the derivation of the constitutive equation.

By substituting for $d\bar{\Lambda}$ from equation (7.43) into equation (7.41), one derives the general inelastic constitutive equation in the configuration $\bar{C}$ as follows:

$$d\bar{\sigma}_{ij} = \bar{D}_{ijkl}\bar{d}_{kl} \tag{7.66}$$

where the elasto-plastic stiffness tensor $\mathbf{D}$ is given by:

$$\bar{D}_{ijkl} = \bar{E}_{ijkl} - \frac{1}{\bar{Q}}\frac{\partial f}{\partial \bar{\sigma}'_{mn}}\bar{E}_{mnkl}\bar{E}_{ijpq}\frac{\partial f}{\partial \bar{\sigma}_{pq}} \tag{7.67}$$

The next step is to use the transformation equations developed in the previous sections in order to obtain a constitutive equation in the configuration C similar to equation (7.66).

7.3.3   Coupling of Damage and Plastic Deformation

In this section, the transformation equations developed in sections 7.1 and 7.2 are used with the constitutive model of the previous section in order to transform the inelastic constitutive equation (7.66) in the configuration $\bar{C}$ to a general constitutive equation in the configuration C that accounts for both damage and plastic deformation.

Using equation (7.28) and taking its derivative, one obtains the following transformation equation for $\bar{\mathbf{d}}'$:

$$\bar{d}'_{ij} = dM^{-T}_{ijmn}\epsilon'_{mn} + M^{-T}_{ijkl}d'_{kl} \tag{7.68}$$

where dM^{-T} is obtained by taking the derivative of the identity $\mathbf{M}^T\mathbf{M}^{-T} = \mathbf{I}$ and noting that $d\mathbf{I} = \mathbf{0}$. Thus, one has:

$$dM_{klmn}^{-T} = -M_{ijkl}^{-T}dM_{ijpq}^{T}M_{pqmn}^{-T} \tag{7.69}$$

The derivative $d\mathbf{M}$ is obtained by using the chain rule as follows:

$$dM_{ijpq} = \frac{\partial M_{ijpq}}{\partial \phi_{mn}}d\phi_{mn} \tag{7.70a}$$

Also, the corotational derivative $\overset{\circ}{\mathbf{M}}$ may be used defined by the following Lie derivative given by Oldroyd (1950):

$$\overset{\circ}{M}_{ijmn} = dM_{ijmn} - \Omega_{ip}M_{pjmn} - \Omega_{jq}M_{iqmn} - \Omega_{mr}M_{ijrn} - \Omega_{ns}M_{ijms} \tag{7.70b}$$

The transformation equation (7.68) for the effective elastic strain rate tensor $\bar{\mathbf{d}}'$ represents a nonlinear relation unlike that of the effective stress tensor of equation (7.1). A similar nonnlinear transformation equation (7.52) was previously derived for the effective plastic strain $\mathbf{d}''$. These two equations will now be used in the derivation of the constitutive model.

Now we are ready to derive the inelastic constitutive relation in the configuration C. Starting with the constitutive equation (7.66) and substituting for $\bar{\sigma}_{ij}$ and $\bar{d}_{kl}$ from equations (7.1) and (7.18), respectively, along with equations (7.68) and (7.52), we obtain:

$$dM_{ijmn}\sigma_{mn}+M_{ijpq}d\sigma_{pq} = \bar{D}_{ijkl}(M_{klmn}^{-T}d'_{mn}+dM_{klpq}^{-T}\epsilon'_{pq}+X_{klmn}d''_{mn}+Z_{kl}) \tag{7.71}$$

Next, we substitute for $d\mathbf{M}$ and $d\mathbf{M}^{-\mathbf{T}}$ from equations (7.70) and (7.69), respectively, for $\mathbf{d}''$ from equation (7.18), and for $\mathbf{d}'$ from a similar equation to (7.40), i.e., $d'_{ij} = E_{ijkl}^{-1}d\sigma_{kl}$, into equation (7.71), the resulting expression is:

$$\frac{\partial M_{ijmn}}{\partial \phi_{pq}}d\phi_{pq}\sigma_{mn} + M_{ijpq}d\sigma_{pq} = \bar{D}_{ijkl}(M_{klmn}^{-T}E_{pqmn}^{-1}d\sigma_{pq}-$$

$$M_{xykl}^{-T}\frac{\partial M_{xyuv}^{T}}{\partial \phi_{mn}}d\phi_{mn}M_{uvpq}E_{cdpq}^{-1}\sigma_{cd} + X_{klmn}d_{mn} - X_{klmn}E_{abmn}^{-1}d\sigma_{ab} + Z_{kl}) \tag{7.72}$$

Finally, we substitute for $d\phi$ from equation (7.63) into equation (7.72) and solve for $d\sigma$ in terms of $\mathbf{d}$. After several algebraic manipulations, we obtain the desired inelastic constitutive relation in the configuration C as follows:

$$d\sigma_{ij} = D_{ijkl}d_{kl} + G_{ij} \tag{7.73}$$

where the effective elasto-plastic stiffness tensor **D** and the additional tensor **G** (comparable to the plastic relaxation stress introduced by Simo and Ju (1987)) are given by:

$$D_{ijkl} = O^{-1}_{pqij}\bar{D}_{pqmn}X_{mnkl} \tag{7.74a}$$

$$G_{ij} = O^{-1}_{pqij}\bar{D}_{pqmn}Z_{mn} \tag{7.74b}$$

where the fourth-order tensor **O** is given by:

$$O_{ijpq} = M_{ijpq} + \frac{\frac{\partial M_{ijmn}}{\partial \phi_{uv}}\frac{\partial g}{\partial \sigma_{pq}}\frac{\partial g}{\partial \sigma_{uv}}\sigma_{mn}}{\frac{\partial L}{\partial l} - \frac{\partial g}{\partial \phi_{xy}}\frac{\partial g}{\partial \sigma_{xy}}} - \bar{D}_{ijkl}(M^{-T}_{klmn}E^{-1}_{pqmn} - X_{klmn}E^{-1}_{pqmn}$$

$$-M^{-T}_{xykl}\frac{\partial M^{-T}_{xyuv}}{\partial \phi_{mn}}\frac{\frac{\partial g}{\partial \sigma_{pq}}\frac{\partial g}{\partial \sigma_{mn}}}{\frac{\partial L}{\partial l} - \frac{\partial g}{\partial \phi_{ab}}\frac{\partial g}{\partial \sigma_{ab}}}M^{-T}_{uvcd}M^{-T}_{rscd}\sigma_{rs}) \tag{7.75}$$

The effective elasto-plastic stiffness tensor **D** in equation (7.74a) is the stiffness tensor including the effects of damage and plastic deformation. It is derived in the configuration of the deformed and damaged body. Equations (7.74) can now be used in finite element analysis. However, it should be noted that the constitutive relation in equation (7.73) represents a nonlinear transformation that makes the numerical implementation of this model impractical. This is due to the additional term G_{ij} which can be considered as some residual stress due to the damaging process. Nevertheless, the constitutive equation becomes linear provided that $G_{ij} = 0$. This is possible only when the term $(\sigma_{mn} - X_{mn})N_{ijrr}$ vanishes as seen in equations (7.74b) and (7.54) and therefore:

$$d\sigma_{ij} = D_{ijkl}d_{kl} \tag{7.76}$$

Upon investigation of the nonlinear constitutive equation (7.73), it is seen that the extra term G_{ij} is due to the linear transformation of the effective stress σ and $\bar{\sigma}$ in equation (7.1). It was shown in equation (7.4) that this transformation leads to a nonlinear relation between σ' and $\bar{\sigma}'$. The authors have shown in a recent work (Voyiadjis and Kattan, 1999) that a linear constitutive equation similar to equation (7.76) can be obtained if a linear transformation is assumed between the deviatoric stresses σ' and $\bar{\sigma}'$ in the form $\bar{\sigma}'_{ij} = M_{ijkl}\sigma'_{kl}$.

For completeness, one can obtain an identity that may be helpful in the numerical calculations. This is done by using the plastic volumetric incompressibility condition (which results directly from equation (7.34)):

$$\bar{d}''_{kk} = 0 \tag{7.77}$$

in the configuration $\bar{C}$. Equation (7.77) is commonly used in metal plasticity without damage (Paulun and Pechervski, 1985; Voyiadjis ans Kattan, 1989). Using equation (7.34) along with the condition (7.77), one obtains the useful identity:

$$N_{rrkl}(\sigma^{\circ}_{kl} - X_{kl}) = 0 \tag{7.78}$$

Equation (7.78) is consistent with the previous conclusion of equation (7.7) since it was shown earlier that $N_{rrkl} = 0$.

In finite element calculations the critical state of damage is reached when the overall parameter l reaches a critical value called l_{cr} in at least one of the element. This value determines the initiation of microcracks and other damaging defects. Alternatively, one can assign several critical values $l_{cr}^{(1)}$, $l_{cr}^{(2)}$, etc. for different damage effects. In order to determine these critical values, which may be considered as material parameters, a series of uniaxial extension tests are to be performed on tensile specimens and the stress-strain curves drawn.

In order to determine $l_{cr}^{(i)}$ (the value of l at which damage initiation starts for a particular damage process "i"), the tensile specimen has to be sectioned at each load increment. The cross-section is to be examined for any cracks or cavities. The load step when cracks first appear in terms of the strain ϵ_1 is to be recorded and compared with the graph of l vs ϵ_1. The corresponding value of l obtained in this way will be taken to be the critical value for l_{cr}. This value is to be used in the finite element analysis of more complicated problems. For more details, see the papers by Chow and Wang (1988) and Voyiadjis (1984, 1988).

7.4   Application to Void Growth: Gurson's Model

Gurson (1977) proposed a yield function $f(\sigma, v)$ for a porous solid with a randomly distributed volume fraction of voids. Gurson's model was used later (Tvergaard, 1982; Tvergaard and Needleman, 1984) to study necking and failure of damaged solids. Tvergaard and Needleman (1984) modified Gurson's yield function in order to account for rate sensitivity and necking instabilities in plastically deforming solids. The modified yield function is used here in the following form (which includes kinematic hardening):

$$f = (\sigma'_{ij} - X'_{ij})(\sigma'_{ij} - X'_{ij}) + 2q_1\sigma_F^2 v \, cosh(\frac{\sigma_{kk}}{2\sigma_F}) - \sigma_F^2(1 + q_2v^2) = 0 \quad (7.79)$$

where σ_F is the yield strength of the matrix material and q_1 and q_2 are material parameters introduced by Tvergaard (1982) to improve agreement between Gurson's model and other results. In equation (7.79), the variable v denotes the void volume fraction in the damaged material. In Gurson's model, damage is characterized by void growth only. The void growth is described by the rate of change of v given by (Voyiadjis, 1988):

$$dv = (1 - \nu)d''_{kk} \quad (7.80)$$

In Gurson's model, it is assumed that the voids remain spherical in shape through the whole process of deformation and damage. The change of shape of voids, their coalescence and nucleation of new voids are ignored in the model. Equation (7.80) implies also that the plastic volumetric change, d''_{kk}, does not vanish for a material with voids.

In the following, it is shown how the proposed model outlined in the first sections of this chapter can be used to obtain the damage effect tensor **M** as applied to Gurson's yield function. It is also shown how certain expressions can be derived for the parameters q_1 and q_2 in a consistent manner. One first starts with the yield function f in the configuration C. Therefore, using equation (7.29) in the form:

$$f = (\bar{\sigma}'_{mn} - \bar{X}'_{mn})(\bar{\sigma}'_{mn} - \bar{X}'_{mn}) - \frac{2}{3}\bar{\sigma}_o^2 \quad (7.81)$$

where the term $c\bar{\kappa}$ is dropped since isotropic hardening is not displayed by Gurson's function. Using the transformation equations (7.10), (7.16), and (7.17) and noting that $\sigma_F^2 = 2\bar{\sigma}_o^2/3$, equation (7.81) becomes:

$$f = H_{ijkl}(\sigma_{ij} - X_{ij})(\sigma_{kl} - X_{kl}) - \sigma_F^2 \quad (7.82)$$

It is noticed that equation (7.81) corresponds exactly to Gurson's function of equation (7.79) with $v = 0$. Using equation (7.2) and (7.12) to transform the total stresses in equation (7.82) into deviatoric stresses, one obtains:

$$f = H_{ijkl}(\sigma'_{ij} - X'_{ij})(\sigma'_{kl} - X'_{kl}) + \frac{1}{9}H_{mmnn}(\sigma_{pp} - X_{qq})^2 - \sigma_F^2 \quad (7.83)$$

Equation (7.83) represents the yield function f in the configuration C, which can now be compared to Gurson's yield function of equation (7.79).

Thus, upon comparing equation (7.79) with equation (7.83) , it is clear that the deviatoric parts of the two functions must be equal. Therefore, one obtains:

$$H_{ijkl}(\sigma'_{ij} - X'_{ij})(\sigma'_{kl} - X'_{kl}) = (\sigma'_{rs} - X'_{rs})(\sigma'_{rs} - X'_{rs}) \qquad (7.84)$$

On the other hand, upon equating the remaining parts of the two functions, one obtains:

$$\frac{1}{9}H_{mmnn}(\sigma_{pp} - X_{qq})^2 = 2q_1\sigma_F^2 v \, cosh(\frac{\sigma_{kk}}{2\sigma_F}) - q_2\sigma_F^2 v^2 \qquad (7.85)$$

The problem is now reduced to manipulating equations (7.84) and (7.85). Rewriting equation (7.84) in the following form:

$$(H_{ijkl} - \delta_{ik}\delta_{jl})(\sigma'_{ij} - X'_{ij})(\sigma'_{kl} - X'_{kl}) = 0 \qquad (7.86)$$

One concludes that the tensor **H** is constant for Gurson's model and can be expressed as follows:

$$H_{ijkl} = \delta_{ik}\delta_{jl} \qquad (7.87)$$

It is clear that the deviatoric part of Gurson's yield function does not display any damage characteristics as given by equation (7.79). This is further supported by equation (7.87) where the damage effect tensor is independent of the damage variable ϕ. Next, upon considering equation (7.87), one obtains $H_{mmnn} = 3$. Substituting this value into equation (7.85) yields:

$$\frac{1}{3}(\sigma_{pp} - X_{qq})^2 = [2q_1cosh(\frac{\sigma_{kk}}{2\sigma_F}) - q_2v]v\sigma_F^2 \qquad (7.88)$$

Equation (7.88) must be satisfied for a possible relationship between Gurson's model and the proposal model. Equation (7.88), as it stands, does not seem to merit an explicit relationship between parameters q_1, q_2, and v. This is due to the presence of the "cosh" term on the right-hand side. Therefore, it is clear that one cannot proceed further without making some assumptions. In particular, two assumptions are to be employed. The first assumption is valid for small values of $\sigma_{kk}/(2\sigma_F)$, where only the first two terms in the "cosh" series expansion are considered:

$$cosh(\frac{\sigma_{kk}}{2\sigma_F}) = 1 + \frac{\sigma_{kk}^2}{8\sigma_F^2} \qquad (7.89)$$

The second assumption concerns the term X_{qq} which appears in the equation (7.88). For the following to be valid, one needs to consider a modified Gurson yield function where the volumetric stress σ_{kk} is replaced by $(\sigma_{kk} - X_{qq})$. Therefore, upon incorporating the above two assumptions into equation (7.88), one obtains:

$$\frac{1}{3}(\sigma_{kk} - X_{qq})^2 = \frac{1}{4}q_1 v(\sigma_{kk} - X_{qq})^2 + (2q_1 - q_2 v)v\sigma_F^2 \qquad (7.90)$$

It is now clear from the equation (7.90) that the following two expressions for q_1 and q_2 in terms of v, need to be satisfied:

$$q_1 = \frac{4}{3v} \qquad (7.91\text{a})$$

$$q_2 = \frac{8}{3v^2} \qquad (7.91\text{b})$$

The relations (7.91a,b) represent variable expressions for the parameters q_1 and q_2 in terms of the void volume fraction, in contrast to the constant values that were suggested earlier by various authors. The relations (7.91a,b) are consistently derived and although they are approximate, in the authors' opinion, they form a basis for more sophisticated expressions. In addition, they are based on a solid derivation which cannot be said for the constant values that were used in the literature. Finally, one more important point that came up in the derivation needs to be considered. As it stands, Gurson's function of equation (7.79) cannot be related to the work presented here. It is a modified version of Gurson's function containing the term $cosh[(\sigma_{kk} - X_{qq})/(2\sigma_F)]$ instead of $cosh[\sigma_{kk}/(2\sigma_F)]$ that is used in the derivation of equations (7.91a,b). The authors believe that this point should be pursued and the proposed modified Gurson function explored further.

7.5 Effective Spin Tensor

In this section, a formal derivation is presented for the transformation equation of the modified spin tensor that is used in the corotational rate equations. In the configuration $\bar{C}$, the corotational derivative of the effective Cauchy stress tensor is given by:

$$\overset{\circ}{\bar{\sigma}}_{ij} = d\bar{\sigma}_{ij} - \bar{\Omega}_{ip}\bar{\sigma}_{pj} + \bar{\sigma}_{iq}\bar{\Omega}_{qj} \qquad (7.92)$$

where $\bar{\Omega}$ is the effective modified spin tensor. The problem now reduces to finding a relation between Ω and $\bar{\Omega}$. One should keep in mind, however, that equation (7.92) is valid only when a Cartesian coordinate system is used. The same remark applies to equation (7.64) in the configuration C.

In order to derive the required relation, one first starts with the transformation equation (7.1). Taking the corotational derivative of this equation and rearranging the terms, one obtains:

$$\overset{\circ}{\sigma}_{ke} = M_{ijke}^{-1}\left(\overset{\circ}{\bar{\sigma}}_{ij} - \overset{\circ}{M}_{ijmn}\sigma_{mn}\right) \tag{7.93}$$

Substituting for $\overset{\circ}{\bar{\sigma}}_{ij}$ from equation (7.92) into equation (7.93) and using the material time derivative $d\sigma_{ij} = dM_{ijrs}\sigma_{rs} + M_{ijpq}d\sigma_{pq}$, one obtains:

$$\overset{\circ}{\sigma}_{kl} = M_{ijkl}^{-1}dM_{ijrs}\sigma_{rs} + d\sigma_{kl} - M_{ijkl}^{-1}\bar{\Omega}_{ip}M_{pjab}\sigma_{ab} + M_{ijkl}^{-1}M_{iqcd}\sigma_{cd}\bar{\Omega}_{qj}$$
$$- M_{ijkl}^{-1}M_{ijmn}\sigma_{mn} \tag{7.94}$$

Comparing the two corotational derivatives appearing in equations (7.64) and (7.94), and after performing some tedious algebraic manipulations, one can finally obtain a relation between $\boldsymbol{\Omega}$ and $\bar{\boldsymbol{\Omega}}$ in the following form:

$$\bar{\Omega}_{mn} = A_{mnpq}\Omega_{pq} + B_{mn} \tag{7.95}$$

where the tensors $\mathbf{A}$ and $\mathbf{B}$ are given by:

$$A_{mnpq} = -C_{mnkl}^{-1}\left(\delta_{kp}\sigma_{ql} - \delta_{lq}\sigma_{kp}\right) \tag{7.96a}$$

$$B_{mn} = -C_{mnkl}^{-1}M_{pqkl}^{-1}\left(dM_{pqrs} - \overset{\circ}{M}_{pqrs}\right)\sigma_{rs} \tag{7.96b}$$

and the tensor $\mathbf{C}$ is given by:

$$C_{klip} = \left(M_{qpkl}^{-1}M_{qicd} - M_{ijkl}^{-1}M_{pjcd}\right)\sigma_{cd} \tag{7.96c}$$

With the availability of the transformation equation of the spin tensor, the theory presented in this chapter is now complete.

Problems

7.1. Explain why equation (7.4) does not represent a linear transformation?

7.2. Derive equation (7.6) explicitly.

7.3. Derive equations (7.8) and (7.9) explicitly.

7.4. Derive equation (7.11) explicitly.

7.5. Derive equations (7.27) and (7.28) explicitly.

7.6. Assume a linear transformation in the form $\bar{\sigma}'_{ij} = M_{ijkl}\sigma'_{kl}$. Derive the corresponding elasto-plastic constitutive model for this case. Specifically, derive a new elasto-plastic stiffness tensor D_{ijkl}.

7.7. Derive equations (7.34) and (7.35) explicitly.

7.8. Derive equation (7.43) explicitly.

7.9. Derive equation (7.45) explicitly.

7.10. Derive equations (7.50) and (7.52) explicitly.

7.11. Consider a von Mises type yield function of the form given by equation (7.29). Neglect isotropic hardening and derive a new elasto-plastic constitutive model based on the modified yield criterion.

7.12. Consider a von Mises type yield function of the form given by equation (7.29). Neglect kinematic hardening and derive a new elasto-plastic constitutive model based on the modified yield criterion.

7.13. Consider a von Mises type yield function of the form given by equation (7.29). Neglect both isotropic and kinematic hardening, and derive a new elasto-plastic constitutive model based on the modified yield criterion.

7.14. Explore using a Tresca type yield function instead of a von Mises in the formulation of the elasto-plastic constitutive model given in this chapter.

7.15. Consider using a different damage evolution criterion by introducing the generalized force Y_{ij} associated with ϕ_{ij} in the equation (7.55). Derive a new evolution equation for the damage variable ϕ in this case.

7.16. Derive equations (7.62) and (7.63) explicitly.

7.17. Derive equation (7.67) explicitly.

7.18. Derive equation (7.70b) explicitly.

7.19. Derive equations (7.73), (7.74) and (7.75) explicitly.

7.20. Explain why there is an additional term G_{ij} in the elasto-plastic constitutive equation (7.73). What are the conditions under which this term vanishes.

7.21. Derive equations (7.87) and (7.88) explicitly.

7.22. Derive equation (7.89) explicitly.

7.23. Derive equation (7.91a,b) explicitly.

7.24. Derive equations (7.95) and (7.96) explicitly.

8

Kinematics of Damage for Finite-Strain Elasto-Plastic Solids

In this chapter, the kinematics of damage for finite strain elasto-plastic deformation is introduced using the fourth-order damage effect tensor through the concept of the effective stress within the framework of continuum damage mechanics. In the absence of the kinematic description of damage, deformation leads one to adopt one of the following two different hypotheses for the small deformation problems. One uses either the hypothesis of strain equivalence or the hypothesis of energy equivalence in order to characterize the damage in the material. The proposed approach in this chapter provides a general description of kinematics of damage applicable to finite strains. This is accomplished by directly considering the kinematics of the deformation field and, furthermore, noting that it is not confined to small strains as in the case of the strain equivalence or the strain energy equivalence approaches. In this chapter, damage is described kinematically in both the elastic domain and the plastic domain using the fourth-order damage effect tensor, which is a function of the second-order damage tensor. The damage effect tensor is explicitly characterized in terms of a kinematic measure of damage through a second-order damage tensor. Two kinds of second-order damage tensor representations are used with respect to two reference configurations. The finite elasto-plastic deformation behavior with damage is also viewed here within the framework of thermodynamics with internal state variables. Using the consistent thermodynamic formulation, one introduces separately the strain due to damage and the associated dissipation energy due to this strain.

8.1 Theoretical Preliminaries

A continuous body in an initial undeformed configuration that consists of the material volume Ω^o is denoted by C^o, while the elasto-plastic damaged deformed configuration at time t after the body is subjected to a set of external agencies is denoted by C^t. The corresponding material volume at time t is denoted by Ω^t. Upon elastic unloading from the configuration C^t an intermediate stress-free configuration is denoted by C^{dp}. In the framework of continuum damage mechanics, a number of fictitious configurations, based on the effective stress concept, are assumed that are obtained by fictitiously removing all the damage that the body has undergone. Thus, the fictitious configuration of the body denoted by $\bar{C}^t$ is obtained from C^t by fictitiously removing all the damage that the body has undergone at C^t. Also, the fictitious configuration denoted by $\bar{C}^p$ is assumed which is obtained from C^{dp} by fictitiously removing all the damage that the body has undergone at C^{dp}. While the configuration $\bar{C}^p$ is the intermediate configuration upon unloading from the configuration $\bar{C}^t$, the initial undeformed body may have a pre-existing damage state. The initial fictitious effective configuration denoted by $\bar{C}^o$ is defined by removing the initial damage from the initial undeformed configuration of the body. In the case of no initial damage existing in the undeformed body, the initial fictitious effective configuration is identical to the initial undeformed configuration. Cartesian tensors are used in this chapter and the tensorial index notation is employed in all equations. The tensors used in the text are denoted by boldface letters. However, superscripts in the notation do not indicate tensorial indices but merely stand for the corresponding deformation configurations such as "e" for elastic, "p" for plastic, and "d" for damage, etc. The barred and tilded notations refer to the fictitious effective configurations.

8.2   Description of Damage State

The damage state can be described using an even order tensor (Leckie, 1993; Onat, 1986; Betten, 1986). Ju (1990) pointed out that even for isotropic damage one should employ a damage tensor (not a scalar damage variable) to characterize the state of damage in materials. However, the damage generally is anisotropic due to the external agency condition or the material nature itself. Although the fourth-order damage tensor can be used directly as a linear transformation tensor to define the effective stress tensor, it is not easy to characterize physically the fourth-order damage tensor compared with the second-order damage tensor. In this chapter, the damage is considered as a symmetric second-order tensor. However, the damage tensor for finite elasto-plastic deformation can be defined in two reference systems (Murakami, 1988). The first one is the damage tensor denoted by ϕ representing the damage state with respect to the current damage configuration C^t. Another one is denoted by $\bar{\phi}$ and is representing the damage state with respect to the elastically unloaded damage configuration C^{dp} (see Chapter 3). Both are given by Murakami (1983) as follows:

$$\phi_{ij} = \sum_{k=1}^{3} \hat{\phi}_k \hat{n}_i^k \hat{n}_j^k \text{ (no sum over } k) \tag{8.1}$$

$$\bar{\phi}_{ij} = \sum_{k=1}^{3} \hat{\bar{\phi}}_k \hat{m}_i^k \hat{m}_j^k \text{ (no sum over } k) \tag{8.2}$$

where $\hat{\mathbf{n}}^k$ and $\hat{\mathbf{m}}^k$ are eigenvectors corresponding to the eigenvalues, $\hat{\phi}_k$ and $\hat{\bar{\phi}}_k$, of the damage tensors ϕ and $\bar{\phi}$, respectively. Equations (8.1) and (8.2) can be written alternately as follows:

$$\phi_{ij} = b_{ir} b_{js} \hat{\phi}_{rs} \tag{8.3}$$

$$\bar{\phi}_{ij} = c_{ir} c_{js} \hat{\bar{\phi}}_{rs} \tag{8.4}$$

The damage tensors in the coordinate system that coincides with the three orthogonal principal directions of the damage tensors $\hat{\phi}_{rs}$ and $\hat{\bar{\phi}}_{rs}$, in equations (8.3) and (8.4), are obviously of diagonal form and are given by:

$$\hat{\phi}_{ij} = \begin{pmatrix} \hat{\phi}_1 & 0 & 0 \\ 0 & \hat{\phi}_2 & 0 \\ 0 & 0 & \hat{\phi}_3 \end{pmatrix} \tag{8.5}$$

$$\hat{\bar{\phi}}_{ij} = \begin{pmatrix} \hat{\bar{\phi}}_1 & 0 & 0 \\ 0 & \hat{\bar{\phi}}_2 & 0 \\ 0 & 0 & \hat{\bar{\phi}}_3 \end{pmatrix} \tag{8.6}$$

and the second-order transformation tensors **b** and **c** are given by:

$$b_{ir} = \begin{pmatrix} n_1^1 & n_2^1 & n_3^1 \\ n_1^2 & n_2^2 & n_3^2 \\ n_1^3 & n_2^3 & n_3^3 \end{pmatrix} \tag{8.7}$$

$$c_{ir} = \begin{pmatrix} m_1^1 & m_2^1 & m_3^1 \\ m_1^2 & m_2^2 & m_3^2 \\ m_1^3 & m_2^3 & m_3^3 \end{pmatrix} \tag{8.8}$$

These proper orthogonal transformation tensors require that:

$$b_{ij} b_{kj} = c_{ij} c_{kj} = \delta_{ik} \tag{8.9}$$

where δ_{ik} is the Kronecker delta and the determinants of the matrices [b] and [c] are given by:

$$|b| = |c| = 1 \qquad (8.10)$$

The relation between the damage tensors ϕ and $\bar{\phi}$ is shown in section 8.4.

8.3 Fourth-Order Anisotropic Damage Effect Tensor

In a general state of deformation and damage, the effective stress tensor $\bar{\sigma}$ is related to the Cauchy stress tensor σ by the following linear transformation as was shown in previous chapters (Murakami and Ohno, 1981):

$$\bar{\sigma}_{ij} = M_{ijkl}\sigma_{kl} \qquad (8.11)$$

where $\mathbf{M}$ is a fourth-order linear transformation operator called the damage effect tensor. Depending on the form used for $\mathbf{M}$, it is very clear from equation (8.11) that the effective stress tensor $\bar{\sigma}$ is generally nonsymmetric. Using a nonsymmetric effective stress tensor as given by equation (8.11) to formulate a constitutive model will result in the introduction of the Cosserat and micropolar continua. However, the use of such complicated mechanics can be easily avoided if the proper fourth-order linear transformation tensor is formulated in order to symmetrize the effective stress tensor. Such a linear transformation tensor called the damage effect tensor is obtained in the literature (Lee et al., 1986; Cordebois and Sidoroff, 1979) using symmetrization methods. One of the symmetrization methods given by Cordebois and Sidoroff (1979) and Lee et al. (1986) is expressed as follows:

$$\bar{\sigma}_{ij} = (\delta_{ik} - \phi_{ik})^{-1/2}\sigma_{kl}(\delta_{jl} - \phi_{jl})^{-1/2} \qquad (8.12)$$

The fourth-order damage effect tensor corresponding to equation (8.12) is defined such that:

$$M_{ijkl} = (\delta_{ik} - \phi_{ik})^{-1/2}(\delta_{jl} - \phi_{jl})^{-1/2} \qquad (8.13)$$

In order to describe the kinematics of damage, the physical meaning of the fourth-order damage effect tensor should be interpreted and not merely given as the symmetrization of the effective stress. In this chapter, the fourth-order damage effect tensor given by equation (8.13) will be used because of its geometrical symmetrization of the effective stress (Corde-bois and Sidoroff, 1979). However, it is very difficult to obtain the explicit representation of $(\delta_{ik} - \phi_{ik})^{-1/2}$. The explicit representation of the fourth-order damage effect tensor **M** using the second-order damage tensor ϕ is of particular importance in the implementation of the constitutive modeling of damage mechanics. Therefore, the damage effect tensor **M** of equation (8.13) should be obtained using the coordinate transformation of the principal damage direction coordinate system. Thus, the fourth-order damage effect tensor given by equation (8.13) can be written as follows (Voyiadjis and Park, 1996a,b).

$$M_{ikjl} = b_{mi}b_{nj}b_{pk}b_{ql}\hat{M}_{mnpq} \tag{8.14}$$

where $\hat{\mathbf{M}}$ is a fourth-order damage effect tensor with the reference to the principal damage direction coordinate system. The fourth-order damage effect tensor $\hat{\mathbf{M}}$ can be written as follows (Voyiadjis and Park, 1996a,b):

$$\hat{M}_{mnpq} = \hat{a}_{mp}\hat{a}_{nq} \tag{8.15}$$

where the second order tensor $\hat{\mathbf{a}}$ in the principal damage direction coordinate system is given by:

$$\hat{a}_{mp} = (\delta_{mp} - \hat{\phi}_{mp})^{-1/2} \equiv \begin{bmatrix} \dfrac{1}{\sqrt{1-\hat{\phi}_1}} & 0 & 0 \\ 0 & \dfrac{1}{\sqrt{1-\hat{\phi}_2}} & 0 \\ 0 & 0 & \dfrac{1}{\sqrt{1-\hat{\phi}_3}} \end{bmatrix} \tag{8.16}$$

Substituting equation (8.15) into equation (8.14), one obtains the following relation:

$$M_{ijkl} = b_{mi}b_{nj}b_{pk}b_{ql}\hat{a}_{mp}\hat{a}_{nq} = a_{ik}a_{jl} \tag{8.17}$$

Using equation (8.17), a second-order tensor **a** is defined as follows:

$$a_{ik} = b_{mi}b_{pk}\hat{a}_{mp} \tag{8.18}$$

The matrix form of equation (8.18) is given as follows (Voyiadjis and Park, 1995a,b):

$$[\mathbf{a}] = [\mathbf{b}]^T [\hat{\mathbf{a}}] [\mathbf{b}]$$

$$= \begin{bmatrix} \dfrac{b_{11}b_{11}}{\sqrt{1-\hat{\phi}_1}} + \dfrac{b_{21}b_{21}}{\sqrt{1-\hat{\phi}_2}} + \dfrac{b_{31}b_{31}}{\sqrt{1-\hat{\phi}_3}} & \dfrac{b_{11}b_{12}}{\sqrt{1-\hat{\phi}_1}} + \dfrac{b_{21}b_{22}}{\sqrt{1-\hat{\phi}_2}} + \dfrac{b_{31}b_{32}}{\sqrt{1-\hat{\phi}_3}} \\[2mm] \dfrac{b_{21}b_{11}}{\sqrt{1-\hat{\phi}_1}} + \dfrac{b_{22}b_{21}}{\sqrt{1-\hat{\phi}_2}} + \dfrac{b_{32}b_{31}}{\sqrt{1-\hat{\phi}_3}} & \dfrac{b_{12}b_{12}}{\sqrt{1-\hat{\phi}_1}} + \dfrac{b_{22}b_{22}}{\sqrt{1-\hat{\phi}_2}} + \dfrac{b_{32}b_{32}}{\sqrt{1-\hat{\phi}_3}} \\[2mm] \dfrac{b_{31}b_{11}}{\sqrt{1-\hat{\phi}_1}} + \dfrac{b_{23}b_{21}}{\sqrt{1-\hat{\phi}_2}} + \dfrac{b_{33}b_{31}}{\sqrt{1-\hat{\phi}_3}} & \dfrac{b_{13}b_{12}}{\sqrt{1-\hat{\phi}_1}} + \dfrac{b_{23}b_{22}}{\sqrt{1-\hat{\phi}_2}} + \dfrac{b_{33}b_{32}}{\sqrt{1-\hat{\phi}_3}} \end{bmatrix}$$

$$\begin{bmatrix} \dfrac{b_{11}b_{13}}{\sqrt{1-\hat{\phi}_1}} + \dfrac{b_{21}b_{23}}{\sqrt{1-\hat{\phi}_2}} + \dfrac{b_{31}b_{33}}{\sqrt{1-\hat{\phi}_3}} \\[2mm] \dfrac{b_{12}b_{13}}{\sqrt{1-\hat{\phi}_1}} + \dfrac{b_{22}b_{23}}{\sqrt{1-\hat{\phi}_2}} + \dfrac{b_{32}b_{33}}{\sqrt{1-\hat{\phi}_3}} \\[2mm] \dfrac{b_{13}b_{13}}{\sqrt{1-\hat{\phi}_1}} + \dfrac{b_{23}b_{23}}{\sqrt{1-\hat{\phi}_2}} + \dfrac{b_{33}b_{33}}{\sqrt{1-\hat{\phi}_3}} \end{bmatrix} \tag{8.19}$$

8.4   The Kinematics of Damage for Elasto-Plastic Behavior with Finite Strains

A position of a particle in C^o at t^o is denoted by $\mathbf{X}$ and can be defined at its corresponding position in C^t at t, denoted by $\mathbf{x}$. Furthermore, assuming that the deformation is smooth regardless of damage, one can assume a one-to-one mapping such that:

$$x_k = x_k(\mathbf{X}, t) \tag{8.20}$$

or

$$X_k = X_k(\mathbf{x}, t) \tag{8.21}$$

The corresponding deformation gradient is expressed as follows:

$$F_{ij} = \frac{\partial x_i}{\partial X_j} \tag{8.22}$$

and the change in the squared length of a material filament d$\mathbf{X}$ is used as a measure of deformation such that:

$$(ds)^2 - (dS)^2 = dx_i dx_i - dX_i dX_i \tag{8.23}$$
$$= 2\varepsilon_{ij} dX_i dX_j$$

or

$$(ds)^2 - (dS)^2 = 2\epsilon_{ij}dx_i dx_j \tag{8.24}$$

where $(ds)^2$ and $(dS)^2$ are the squared lengths of the material filaments in the deformed with damage configuration C^t, and the initial undeformed configuration C^o, respectively. ε and ϵ are the Lagrangian and Eulerian strain tensors, respectively, and are given by:

$$\varepsilon_{ij} = \frac{1}{2}(F_{ki}F_{kj} - \delta_{ij}) = \frac{1}{2}(C_{ij} - \delta_{ij}) \tag{8.25}$$

$$\epsilon_{ij} = \frac{1}{2}(\delta_{ij} - F_{ki}^{-1}F_{kj}^{-1}) = \frac{1}{2}(\delta_{ij} - B_{ij}^{-1}) \tag{8.26}$$

where $\mathbf{C}$ and $\mathbf{B}$ are the right Cauchy-Green and the left Cauchy-Green tensors, respectively. The velocity vector field in the current configuration at time t is given by:

$$v_i = \frac{dx_i}{dt} \tag{8.27}$$

The velocity gradient in the current configuration at time t is given by:

$$L_{ij} = \frac{\partial v_i}{\partial x_j} \tag{8.28}$$
$$= dF_{ik}F_{kj}^{-1}$$
$$= D_{ij} + W_{ij}$$

where $d\mathbf{F}$ indicates the material time derivative and where $\mathbf{D}$ and $\mathbf{W}$ are the rate of deformation (stretching) and the vorticity, respectively. The rate of deformation $\mathbf{D}$ is equal to the symmetric part of the velocity gradient $\mathbf{L}$, while the vorticity $\mathbf{W}$ is the antisymmetric part of the velocity gradient $\mathbf{L}$ such that:

$$D_{ij} = \frac{1}{2}(L_{ij} + L_{ji}) \tag{8.29}$$

$$W_{ij} = \frac{1}{2}(L_{ij} - L_{ji}) \tag{8.30}$$

Strain rate measures are obtained by differentiating equations (8.23) and (8.24) such that:

$$\frac{d}{dt}[(ds)^2 - (dS)^2] = 2dX_i d\varepsilon_{ij} dX_j$$
$$= 2dx_i D_{ij} dx_j$$
$$= 2dX_i F_{ik} D_{ij} F_{jm} dX_m$$
$$= 2dx_i(d\epsilon_{ij} + \epsilon_{ik}L_{kj} + L_{ik}\epsilon_{kj})dx_j \tag{8.31}$$

By comparing equations (8.31a) and (8.31c), one obtains the rate of the Lagrangian strain that is the projection of **D** onto the reference frame as follows:

$$d\varepsilon_{ij} = F_{ki} D_{kl} F_{lj} \tag{8.32}$$

while the deformation rate **D** is equal to the Cotter-Rivin connected rate of the Eulerian strain as follows:

$$D_{ij} = d\epsilon_{ij} + \epsilon_{ik} L_{kj} + L_{ik} \epsilon_{kj} \tag{8.33}$$

The convected derivative shown in equation (8.33) can also be interpreted as the Lie derivative of the Eulerian Strain (Lubarda and Krajcinovic, 1995).

8.4.1   A Multiplicative Decomposition

A schematic drawing representing the kinematics of elasto-plastic damage deformation is shown in Figure 8.1. In the figure, C^o is the initial unde-formed configuration of the body which may have initial damage in the material. C^t represents the current elasto-plastically deformed and dam-aged configuration of the body. The configuration of $\bar{C}^o$ represents the initial configuration of the body that is obtained by fictitiously removing the initial damage from the C^o configuration. If the initial configuration is undamaged, consequently there is no difference in the configurations C^o and $\bar{C}^o$. The configuration $\bar{C}^t$ is obtained by fictitiously removing the damage from the configuration C^t. Configuration C^{dp} is an interme-diate configuration upon elastic unloading. In the most general case of large deformation processes, damage may be involved due to void and mi-crocrack development because of external agencies. Although damage at the microlevel is a material discontinuity, damage can be considered as an irreversible deformation process in the framework of continuum damage mechanics. Furthermore, one assumes that upon unloading from the elasto-plastic damage state, the elastic part of the deformation can be completely recovered while no additional plastic deformation and damage takes place. Thus, upon unloading the elasto-plastic damage deformed body from the current configuration C^t will elastically unload to an intermediate stress free configuration denoted by C^{dp} as shown in Figure 8.1. Although the damage process is an irreversible deformation thermodynamically, how-ever, the deformation due to damage itself can be partially or completely recovered upon unloading due to closure of microcracks or contraction of microvoids. Nevertheless, recovery of damage deformation does not mean the healing of damage. The deformation gradient tensor and the Green de-formation tensor of the elasto-plastic damage deformation can be obtained through Path I, Path II, or Path III as shown in Figure 8.1. Consider Path I - the deformation gradient referred to the undeformed configuration C^o, is denoted by **F** and is polarly decomposed into the elastic deformation gradient denoted by **F**^{e} and the damage-plastic deformation gradient denoted by **F**^{dp} such that:

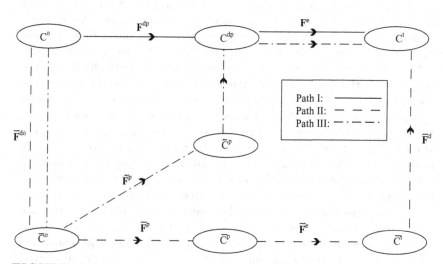

FIGURE 8.1
Schematic representation of elasto-plastic damage deformation configurations.(Voyiadjis and Park, 1999)

$$F_{ij} = F_{ik}^e F_{kj}^{dp}$$
(8.34)

The elastic deformation gradient is given by:

$$F_{ij}^e = \frac{\partial x_i}{\partial x_j^{dp}}$$
(8.35)

The corresponding damage-plastic deformation gradient is given by:

$$F_{ij}^{dp} = \frac{\partial x_i^{dp}}{\partial X_j}$$
(8.36)

The right Cauchy-Green deformation tensor **C** is given by:

$$C_{ij} = F_{nk}^{dp} F_{ki}^e F_{nm}^e F_{mj}^{dp}$$
(8.37)

The finite deformation damage models by Ju (1989) and Zbib (1993) emphasize that "added flexibility" due to the existence of microcracks or microvoids is already embedded in the deformation gradient implicitly. Murakami (1983) presented the kinematics of damage deformation using the second-order damage tensor. However, the lack of an explicit formulation for the kinematics of finite deformation with damage leads to the failure in obtaining an explicit derivation of the kinematics that directly consider the damage deformation. Although most finite strain elasto-plastic deformation processes involve damage such as microvoids, nucleations and microcracks development due to external agencies, however, only the elastic and plastic deformation processes are considered kinematically due to the complexity in the development of damage deformation. In this chapter, the kinematics of damage will be explicitly characterized based on continuum damage mechanics. The elastic deformation gradient corresponds to elastic stretching and rigid body rotations due to both internal and external constraints. The plastic deformation gradient arises from purely irreversible processes due to dislocations in the material. Damage may by initiated and evolves in both the elastic and plastic deformation processes. Particularly, damage in the elastic deformation state is termed elastic damage, which is the case for most brittle materials, while damage in the plastic deformation state is termed plastic damage which is mainly for ductile materials. Additional deformation due to damage consists of damage itself with additional deformation due to elastic and plastic stiffness. In this chapter, kinematics of damage deformation is completely described for both damage and the coupling of damage with elasto-plastic deformation. The total Lagrangian strain tensor is expressed as follows:

$$
\begin{aligned}
\varepsilon_{ij} &= \frac{1}{2}(F_{ki}^{dp} F_{kj}^{dp} - \delta_{ij}) + \frac{1}{2} F_{mi}^{dp}(F_{km}^{e} F_{kn}^{e} - \delta_{mn}) F_{nj}^{dp} \quad (8.38)\\
&= \varepsilon_{ij}^{dp} + F_{mi}^{dp} \epsilon_{mn}^{e} F_{nj}^{dp}\\
&= \varepsilon_{ij}^{dp} + \varepsilon_{ij}^{e}
\end{aligned}
$$

where ε^{dp} and ε^{e} are the Lagrangian damage-plastic strain tensor and the Lagrangian elastic strain tensor, respectively, measured with respect to the reference configuration C^{o}. ϵ^{e} is the Lagrangian elastic strain tensor measured with respect to the intermediate configuration C^{dp}. Similarly, the Eulerian strains corresponding to deformation gradients $\mathbf{F^{e}}$ and $\mathbf{F^{dp}}$ are given by:

$$
e_{ij}^{dp} = \frac{1}{2}(\delta_{ij} - F_{ki}^{dp^{-1}} F_{kj}^{dp^{-1}}) \quad (8.39)
$$

$$
e_{ij}^{e} = \frac{1}{2}(\delta_{ij} - F_{ki}^{e^{-1}} F_{kj}^{e^{-1}}) \quad (8.40)
$$

The Eulerian strain tensor can be expressed as follows:

$$\epsilon_{ij} = \epsilon_{ij}^e + F_{ki}^{e^{-1}} e_{km}^{dp} F_{mj}^{e^{-1}} \tag{8.41}$$
$$= \epsilon_{ij}^e + \epsilon_{ij}^{dp}$$

The strain $\mathbf{e}^{dp}$ is referred to the intermediate configuration C^{dp}, while the strains ϵ, ϵ^e, and ϵ^{dp} are defined relative to the current configuration as a reference. The relationship between the Lagrangian and Eulerian strains is obtained directly in the form:

$$\varepsilon_{ij} = F_{ki} \epsilon_{kl} F_{lj} \tag{8.42}$$

The change in the squared length of a material filament deformed elastically from C^t to C^{dp} is given by:

$$(ds)^2 - (ds^{dp})^2 = dx_i dx_i - dx_i^{dp} dx_i^{dp} \tag{8.43}$$
$$= 2dX_i \varepsilon_{ij}^e dX_j$$

However, the change in the squared length of a material filament due to damage and plastic deformation from C^{dp} to C^o is given by:

$$(ds^{dp})^2 - (dS)^2 = 2dX_i \varepsilon_{ij}^{dp} dX_j \tag{8.44}$$

The kinematics of finite strain elasto-plastic deformation including damage is completely described in Path I. In order to describe the kinematics of damage and plastic deformation, the deformation gradient given by equation (8.34) may be further decomposed into:

$$F_{ij} = F_{ik}^e F_{km}^d F_{mj}^p \tag{8.45}$$

However, it is very difficult to characterize physically only the kinematics of deformation due to damage in spite of its obvious physical phenomenon. The damage, however, may be defined through the effective stress concept. Similarly, the kinematics of damage can be described using the kinematic configuration. Considering Path II, the deformation gradient can be alternatively expressed as follows:

$$F_{ij} = \bar{F}_{ik}^d \bar{F}_{km}^e \bar{F}_{mn}^p \bar{F}_{nj}^{do} \tag{8.46}$$

where $\bar{\mathbf{F}}^d$ is the fictitious damage deformation gradient from configuration $\bar{C}^t$ to C^t and is given by:

$$\bar{F}_{ij}^d = \frac{\partial x_i}{\partial \bar{x}_j} \tag{8.47}$$

The elastic deformation gradient in the effective configuration is given by:

$$\bar{F}^e_{ij} = \frac{\partial \bar{x}_i}{\partial \bar{x}^p_j} \tag{8.48}$$

The corresponding plastic deformation gradient in the effective configuration is given by:

$$\bar{F}^p_{ij} = \frac{\partial \bar{x}^p_i}{\partial \bar{X}_j} \tag{8.49}$$

while the fictitious initial damage deformation gradient from configuration $\bar{C}^o$ to C^o is given by:

$$\bar{F}^{do}_{ij} = \frac{\partial \bar{X}_i}{\partial X_j} \tag{8.50}$$

Similar to Path I, the right Cauchy Green deformation tensor, **C**, is given by:

$$C_{ij} = \bar{F}^{do}_{mk} \bar{F}^p_{kp} \bar{F}^e_{pq} \bar{F}^d_{qi} \bar{F}^d_{mn} \bar{F}^e_{nr} \bar{F}^p_{rs} \bar{F}^{do}_{sj} \tag{8.51}$$

The Lagrangian damage strain tensor measured with respect to the fictitious configuration $\bar{C}^t$ is given by:

$$\bar{\epsilon}^d_{ij} = \frac{1}{2} (\bar{F}^d_{ki} \bar{F}^d_{kj} - \delta_{ij}) \tag{8.52}$$

and the corresponding Lagrangian effective elastic strain tensor measured with respect to the fictitious configuration $\bar{C}^p$ is given by:

$$\bar{\epsilon}^e = \frac{1}{2} (\bar{F}^e_{ki} \bar{F}^e_{kj} - \delta_{ij}) \tag{8.53}$$

The Lagrangian effective plastic strain tensor measured with respect to the fictitious undamaged initial configuration $\bar{C}^o$ is given by:

$$\bar{\epsilon}^p_{ij} = \frac{1}{2} (\bar{F}^p_{ki} \bar{F}^p_{kj} - \delta_{ij}) \tag{8.54}$$

The total Lagrangian strain tensor is therefore expressed as follows:

$$
\begin{aligned}
\varepsilon_{ij} = {} & \frac{1}{2} (\bar{F}^{do}_{ki} \bar{F}^{do}_{kj} - \delta_{ij}) + \frac{1}{2} \bar{F}^{do}_{mi} (\bar{F}^p_{km} \bar{F}^p_{kn} - \delta_{mn}) \bar{F}^{do}_{nj} \\
& + \frac{1}{2} \bar{F}^{do}_{ni} \bar{F}^p_{rn} (\bar{F}^e_{qr} \bar{F}^e_{qs} - \delta_{rs}) \bar{F}^p_{sm} \bar{F}^{do}_{mj} \\
& + \frac{1}{2} \bar{F}^{do}_{wi} \bar{F}^p_{nw} \bar{F}^e_{rn} (\bar{F}^d_{qr} \bar{F}^d_{qs} - \delta_{rs}) \bar{F}^e_{sm} \bar{F}^p_{mk} \bar{F}^{do}_{kj}
\end{aligned}
\tag{8.55}
$$

The Lagrangian initial damage strain tensor measured with respect to the reference configuration C^o is denoted by:

$$\bar{\varepsilon}_{ij}^{do} = \frac{1}{2}(\bar{F}_{ki}^{do}\bar{F}_{kj}^{do} - \delta_{ij}) \tag{8.56}$$

The Lagrangian plastic strain tensor measured with respect to the reference configuration C^o is denoted by:

$$\bar{\varepsilon}_{ij}^{p} = \bar{F}_{ki}^{do}\bar{\varepsilon}_{km}^{p}\bar{F}_{mj}^{do} \tag{8.57}$$

One now defines the Lagrangian elastic strain tensor measured with respect to the reference configuration C^o as follows:

$$\bar{\varepsilon}_{ij}^{e} = \bar{F}_{ni}^{do}\bar{F}_{nk}^{p}\bar{\varepsilon}_{km}^{e}\bar{F}_{mr}^{p}\bar{F}_{rj}^{do} \tag{8.58}$$

and the corresponding Lagrangian damage strain tensor measured with respect to the reference configuration C^o is given by:

$$\bar{\varepsilon}_{ij}^{d} = \bar{F}_{wi}^{do}\bar{F}_{wn}^{p}\bar{F}_{nk}^{e}\bar{\varepsilon}_{km}^{d}\bar{F}_{mr}^{e}\bar{F}_{rs}^{p}\bar{F}_{sj}^{do} \tag{8.59}$$

The total Lagrangian strain is now given as follows through the additive decomposition of the corresponding strains:

$$\varepsilon_{ij} = \bar{\varepsilon}_{ij}^{do} + \bar{\varepsilon}_{ij}^{p} + \bar{\varepsilon}_{ij}^{e} + \bar{\varepsilon}_{ij}^{d} \tag{8.60}$$

The change in the squared length of a material filament deformed due to fictitiously removing of damage from C^t to $\bar{C}^t$ is given by:

$$(ds)^2 - (\bar{ds})^2 = dx_i dx_i - d\bar{x}_i d\bar{x}_i$$
$$= 2dX_i\bar{\varepsilon}_{ij}^{d}dX_j \tag{8.61}$$

The change in the squared length of a material filament deformed elastically from $\bar{C}^t$ to $\bar{C}^p$ is given by:

$$(\bar{ds})^2 - (\bar{ds}^P)^2 = d\bar{x}_i d\bar{x}_i - d\bar{x}_i^p d\bar{x}_i^p \tag{8.62}$$
$$= 2dX_i\bar{\varepsilon}_{ij}^{e}dX_j$$

The change in the squared length of a material filament deformed plastically from $\bar{C}^o$ to $\bar{C}^p$ is then given by:

$$(\bar{ds}^P)^2 - (\bar{dS})^2 = d\bar{x}_i^p d\bar{x}_i^p - d\bar{X}_i d\bar{X}_i \tag{8.63}$$
$$= 2\bar{\varepsilon}_{ij}^{p}dX_i dX_j$$

while the change in the squared length of a material filament deformed due to fictitious removing of the initial damage from $\bar{C}^o$ to C^o is given by:

$$(\bar{d}s)^2 - (ds)^2 = d\bar{X}_i d\bar{X}_i - dX_i dX_i \qquad (8.64)$$
$$= 2dX_i \bar{\varepsilon}_{ij}^{do} dX_j$$

Finally, Path III gives the deformation gradient as follows:

$$F_{ij} = F_{il}^e \tilde{F}_{lm}^d \tilde{F}_{mn}^p \bar{F}_{nj}^{do} \qquad (8.65)$$

where $\tilde{\mathbf{F}}^d$ is the fictitious damage deformation gradient from configuration $\tilde{C}^p$ to C^{dp} and is given by:

$$\tilde{F}_{ij}^d = \frac{\partial x_i^{dp}}{\partial \tilde{x}^p} \qquad (8.66)$$

and the corresponding plastic deformation gradient in the effective configuration is given by:

$$\tilde{F}_{ij}^p = \frac{\partial \tilde{x}_i^p}{\partial X_j} \qquad (8.67)$$

Similar to Path II, the Right Cauchy Green deformation tensor $\mathbf{C}$ is given by:

$$C_{ij} = \bar{F}_{mk}^{do} \tilde{F}_{kp}^p \tilde{F}_{pq}^d F_{qi}^e F_{mn}^e \tilde{F}_{nr}^d \tilde{F}_{rs}^p \bar{F}_{sj}^{do} \qquad (8.68)$$

The Lagrangian damage strain tensor measured with respect to the fictitious intermediate configuration $\tilde{C}^p$ is given by:

$$\tilde{\epsilon}_{ij}^d = \frac{1}{2}(\tilde{F}_{ki}^d \tilde{F}_{kj}^d - \delta_{ij}) \qquad (8.69)$$

The total Lagrangian strain tensor is expressed as follows:

$$\bar{\varepsilon}_{ij} = \frac{1}{2}(\bar{F}_{ki}^{do} \bar{F}_{kj}^{do} - \delta_{ij}) + \frac{1}{2}\bar{F}_{mi}^{do}(\tilde{F}_{km}^p \tilde{F}_{kn}^p - \delta_{mn})\bar{F}_{nj}^{do}$$
$$+ \frac{1}{2}\bar{F}_{ni}^{do} \tilde{F}_{rn}^p(\tilde{F}_{qr}^d \tilde{F}_{qs}^d - \delta_{rs})\tilde{F}_{sm}^p \bar{F}_{mj}^{do}$$
$$+ \frac{1}{2}\bar{F}_{wi}^{do} \tilde{F}_{nw}^p \tilde{F}_{rn}^d(F_{qr}^e F_{qs}^e - \delta_{rs})\tilde{F}_{sm}^d \tilde{F}_{mk}^p \bar{F}_{kj}^{do} \qquad (8.70)$$

The Lagrangian damage strain tensor measured with respect to the reference configuration C^o is denoted by:

$$\bar{\varepsilon}_{ij}^d = \bar{F}_{ki}^{do} \tilde{F}_{mk}^p \tilde{\epsilon}_{mn}^d \tilde{F}_{nq}^p \bar{F}_{qj}^{do} \qquad (8.71)$$

The Lagrangian elastic strain tensor measured with respect to the reference configuration C^o is denoted by:

$$\varepsilon_{ij}^e = \bar{F}_{li}^{do} \tilde{F}_{kl}^p \tilde{F}_{mk}^d \epsilon_{mn}^e \tilde{F}_{nq}^d \tilde{F}_{qr}^p \bar{F}_{rj}^{do} \tag{8.72}$$

The corresponding total Lagrangian strain is now given by:

$$\varepsilon_{ij} = \bar{\varepsilon}_{ij}^{do} + \bar{\varepsilon}_{ij}^p + \bar{\varepsilon}_{ij}^d + \bar{\varepsilon}_{ij}^e \tag{8.73}$$

The change in the squared length of a material filament deformed to fictitious removal of damage from C^{dp} to $\tilde{C}^p$ is given by:

$$(ds^{dp})^2 - (d\tilde{s}^p)^2 = dx_i^{dp} dx_i^{dp} - d\tilde{x}_i^p d\tilde{x}_i^p \tag{8.74}$$
$$= 2dx_i \bar{\varepsilon}_{ij}^d dX_j$$

The change in the squared length of a material filament deformed plastically from $\bar{C}^o$ to $\tilde{C}^p$ is then given by:

$$(d\tilde{s}^p)^2 - (d\bar{S})^2 = d\tilde{x}_i^p d\tilde{x}_i^p - d\bar{X}_i d\bar{X}_i \tag{8.75}$$
$$= 2dx_i \bar{\varepsilon}_{ij}^p dX_j$$

The total Lagrangian strain tensors obtained by considering the three paths are given by equations (8.38), (8.60), and (8.73). From the equivalence of these total strains, one obtains the explicit presentations of the kinematics of damage as follows. With the assumption of the equivalence between the elastic strain tensors given by equations (8.38) and (8.73), the damage-plastic deformation gradient given by equation (8.36) and the Lagrangian damage plastic strain tensor can be expressed as follows:

$$F_{ij}^{dp} = \bar{F}_{ik}^{do} \tilde{F}_{kl}^p \tilde{F}_{lj}^d \tag{8.76}$$

and

$$\varepsilon_{ij}^{dp} = \bar{\varepsilon}_{ij}^{do} + \tilde{\varepsilon}_{ij}^p + \tilde{\varepsilon}_{ij}^d \tag{8.77}$$

Furthermore, one obtains the following expression from equations (8.60) and (8.73) as follows:

$$\bar{\varepsilon}_{ij}^e + \tilde{\varepsilon}_{ij}^d = \tilde{\varepsilon}_{ij}^d + \varepsilon_{ij}^e \tag{8.78}$$

which concludes that $\tilde{C}^p$ and $\bar{C}^p$ are the same. Substituting equations (8.59), (8.71), and (8.72) into equation (8.78), one obtains the effective Lagrangian elastic tensor as follows:

$$\bar{\varepsilon}^e_{ij} = \bar{F}^{do}_{ki} \bar{F}^p_{mk} (\tilde{\epsilon}^d_{mn} - \bar{F}^e_{qm} \tilde{\epsilon}^d_{qr} \bar{F}^e_{rn} + \tilde{F}^d_{qm} \epsilon^e_{qr} \tilde{F}^d_{rn}) \bar{F}^p_{ns} \bar{F}^{do}_{sj} \qquad (8.79)$$

Using equations (8.58) and (8.79), one can now express $\bar{\epsilon}$ as follows:

$$\bar{\epsilon}^e_{ij} = \tilde{\epsilon}^d_{ij} - \bar{F}^e_{mi} \tilde{\epsilon}^d_{mn} \bar{F}^e_{nj} + \tilde{F}^d_{mi} \epsilon^e_{mn} \tilde{F}^d_{nj} \qquad (8.80)$$

This expression gives a general relation of the effective elastic strain for finite strains of elasto-plastic damage deformation. For the special case when one assumes that

$$\tilde{\epsilon}^d - \bar{F}^e_{mi} \tilde{\epsilon}^d_{mn} \bar{F}^e_{nj} = 0 \qquad (8.81)$$

equation (8.80) can be reduced to the following expression:

$$\bar{\epsilon}^e_{ij} = \tilde{F}^d_{ki} \epsilon^e_{kl} \tilde{F}^d_{lj} \qquad (8.82)$$

This relation is similar to that obtained without the consideration of the kinematics of damage and only utilizing the hypothesis of elastic energy equivalence. However, equation (8.82) for the case of finite strains is given by relation (8.80) which cannot be obtained through the hypothesis of elastic energy equivalence. Equation (8.81) may be valid only for some special case of the small strain theory.

8.4.2 Fictitious Damage Deformation Gradients

The two fictitious deformation gradients given by equations (8.47) and (8.66) may be used to define the damage tensor in order to describe the damage behavior of solids. Since the fictitious effective deformed configuration denoted by $\bar{C}^t$ is obtained by removing the damages from the real deformed configuration denoted by C^t, therefore, the differential volume of the fictitious effective deformed volumes denoted by $d\bar{\Omega}^t$ is obtained as follows (Onat, 1986):

$$d\bar{\Omega}^t = d\Omega^t - d\Omega^d \qquad (8.83)$$
$$= \sqrt{(1 - \hat{\phi}_1)(1 - \hat{\phi}_2)(1 - \hat{\phi}_3)} \quad d\Omega^t$$

or

$$d\Omega^t = \bar{J}^d d\bar{\Omega}^t \qquad (8.84)$$

where Ω^d is the volume of damage in the configuration C^t and $\bar{J}^d$ is termed the Jacobian of the damage deformation which is the determinant of the fictitious damage deformation gradient. Thus, the Jacobian of the damage deformation can be written as follows:

$$\bar{J}^d = |\bar{F}^d_{ij}| \tag{8.85}$$

$$= \frac{1}{\sqrt{(1 - \hat{\phi}_1)(1 - \hat{\phi}_2)(1 - \hat{\phi}_3)}}$$

The determinant of the matrix [a] in equation (8.19) is given by:

$$|a| = |b|^T |\hat{a}||b| \tag{8.86}$$
$$= |\hat{a}|$$
$$= \frac{1}{\sqrt{(1 - \hat{\phi}_1)(1 - \hat{\phi}_2)(1 - \hat{\phi}_3)}}$$

Thus, one assumes the following relation without loss of generality:

$$\bar{F}^d_{ij} = (\delta_{ij} - \phi_{ij})^{-1/2} \tag{8.87}$$

Although the identity is established between $\bar{J}^d$ and $|a|$, however, this is not sufficient to demonstrate the validity of equation (8.87). This relation is assumed here based on the physics of the geometrically symmetrized effective stress concept (Onat, 1986). Similarly, the fictitious damage deformation gradient $\tilde{\mathbf{F}}^d$ can be written as follows:

$$\tilde{F}^d_{ij} = (\delta_{ij} - \phi_{ij})^{-1/2} \tag{8.88}$$

Finally, assuming that $\bar{\mathbf{x}} = \tilde{\mathbf{x}}$ based on equation (8.78), the relations between $\tilde{\mathbf{F}}^d$ and $\bar{\mathbf{F}}^d$, and ϕ and $\bar{\phi}$ are given by:

$$\tilde{F}^d_{ij} = F^e_{ki} \bar{F}^d_{kl} F^{e\,-1}_{lj} \tag{8.89}$$

$$\bar{\phi}_{ij} = F^e_{ki} \phi_{kl} F^{e\,-1}_{lj} \tag{8.90}$$

8.4.3   An Additive Decomposition

The kinematics of finite deformation is described here based on the polar decomposition by considering three paths as indicated in the previous section. In order to proceed further, one assumes a homogeneous state of deformation such that the completely unloaded stress-free configuration C^{dp} has opened cracks and micro-cavities. Furthermore, one assumes that these cracks and micro-cavities can be completely closed by subjecting them to certain additional stress. The configuration that is subjected to the additional stresses is denoted by C^p and it is assumed that this configuration has only deformed plastically. The additional stress which can close all micro-cracks and micro-cavities is assumed as follows:

$$\sigma_{ij}^* = \sigma_{ij} - \bar{\sigma}_{ij} \tag{8.91}$$

If no initial damage is assumed in the configuration C^o, it can be assumed such that $C^p = \tilde{C}^p$. The total displacement vector $\mathbf{u}(\mathbf{X}, \mathbf{t})$ can be decomposed in the Cartesian reference frame in the absence of rigid body displacement such that:

$$u_i = u_i^e + u_i^d + u_i^p \tag{8.92}$$

$$u_i = x_i - X_i \tag{8.93}$$

$$u_i^e = x_i - x_i^d \tag{8.94}$$

$$u_i^d = x_i^d - x_i^p \tag{8.95}$$

$$u_i^p = x_i^p - X_i \tag{8.96}$$

where $\mathbf{x^d}$ is a point in the intermediate unloaded configuration C^{dp} and $\mathbf{x^P}$ is a point in the configuration $\tilde{C}^p$. Recalling that $\mathbf{u} = \mathbf{x} - \mathbf{X}$ and using the notation $u_{i,j} = \partial x_i / \partial X_j$, the corresponding total Lagrangian strain tensor ε is given by equation (8.25) can be written in the usual form as follows:

$$\varepsilon_{ij} = \frac{1}{2}(u_{i,j} + u_{j,i} + u_{k,i}u_{k,j}) \tag{8.97}$$

Substituting equation (8.92) into equation (8.97), one obtains:

$$\varepsilon_{ij} = \epsilon_{ij}^p + \epsilon_{ij}^d + \epsilon_{ij}^e + \epsilon_{ij}^{de} + \epsilon_{ij}^{pe} + \epsilon_{ij}^{pd} \tag{8.98}$$

where $\boldsymbol{\epsilon}^p$ termed the pure plastic strain is given by:

$$\epsilon_{ij}^p = \frac{1}{2}(u_{i,j}^p + u_{j,i}^p + u_{k,i}^p u_{k,j}^p) \tag{8.99}$$

$\boldsymbol{\epsilon}^d$ termed the pure damage strain is given by:

$$\epsilon_{ij}^d = \frac{1}{2}(u_{i,j}^d + u_{j,i}^d + u_{k,i}^d u_{k,j}^d) \tag{8.100}$$

$\boldsymbol{\epsilon}^e$ termed the pure elastic strain is given by:

$$\epsilon_{ij}^e = \frac{1}{2}(u_{i,j}^e + u_{j,i}^e + u_{k,i}^e u_{k,j}^e) \tag{8.101}$$

ϵ^{de} termed the coupled elastic-damage strain is given by:

$$\epsilon_{ij}^{de} = \frac{1}{2}(u_{k,i}^{e} u_{k,j}^{d} + u_{k,i}^{d} u_{k,j}^{e})$$ (8.102)

ϵ^{pe} termed the coupled elastic-plastic strain is given by:

$$\epsilon_{ij}^{pe} = \frac{1}{2}(u_{k,i}^{e} u_{k,j}^{p} + u_{k,i}^{p} u_{k,j}^{e})$$ (8.103)

and ϵ^{pd} termed the coupled plastic-damage strain is given by:

$$\epsilon_{ij}^{pd} = \frac{1}{2}(u_{k,i}^{p} u_{k,j}^{d} + u_{k,i}^{d} u_{k,j}^{p})$$ (8.104)

One defines the Lagrangian elastic strain as follows:

$$\varepsilon_{ij}^{e} = \epsilon_{ij}^{e} + \epsilon_{ij}^{de} + \epsilon_{ij}^{pe}$$ (8.105)

the Lagrangian damage strain as follows:

$$\varepsilon_{ij}^{d} = \epsilon_{ij}^{d}$$ (8.106)

and the Lagrangian plastic strain as follows:

$$\varepsilon_{ij}^{p} = \epsilon_{ij}^{p} + \epsilon_{ij}^{pd}$$ (8.107)

The coupled term of elastic-damage and plastic-damage strains are linked, respectively, with the elastic and plastic strains since they directly influence the stresses acting on the body. Consequently, the total Lagrangian strain can be written as follows:

$$\varepsilon_{ij} = \varepsilon_{ij}^{p} + \varepsilon_{ij}^{d} + \varepsilon_{ij}^{e}$$ (8.108)

The differential displacement is given by:

$$du_i = x_i^{t+dt} - x_i^{t}$$ (8.109)

Then, the corresponding differential total displacement can be decomposed into an elastic, plastic, and damage parts as follows:

$$du_i = du_i^{e} + du_i^{d} + du_i^{p}$$ (8.110)

Evidently, one obtains the following decomposition of the velocity tensor field $\mathbf{v}(\mathbf{x}, t)$:

$$v_i(x_i, t) = v_i^{e}(x_i, t) + v_i^{d}(x_i, t) + v_i^{p}(x_i, t)$$ (8.111)

where $\mathbf{v}^{e}$ is the velocity vector field due to elastic stretching and rigid body rotations, $\mathbf{v}^{d}$ is the velocity vector field due to the damage process, and

$\mathbf{v}^p$ is the velocity vector field arising from the plastic deformations due to dislocation motion. The gradient of the frame $\mathbf{x}$ is given by the following relation:

$$L_{ij} = L^e_{ij} + L^d_{ij} + L^p_{ij} \tag{8.112}$$

$$D_{ij} = D^e_{ij} + D^d_{ij} + D^p_{ij} \tag{8.113}$$

$$W_{ij} = W^e_{ij} + W^d_{ij} + W^p_{ij} \tag{8.114}$$

8.5   Irreversible Thermodynamics

The finite elasto-plastic deformation behavior with damage can be viewed within the framework of thermodynamics with internal state variables. The Helmholtz free energy per unit mass in an isothermal deformation process at the current state of the deformation and material damage is assumed as follows:

$$\Psi = \psi + \gamma \tag{8.115}$$

where ψ is the strain energy which is a purely reversible stored energy, while γ is the energy associated with specific microstructural changes produced by damage and plastic yielding. Conceptionally, the energy γ is assumed to be an irreversible energy. In general, an explicit presentation of the energy γ and its rate $d\gamma$ is limited by the complexities of the internal microstructural changes, however, only two internal variables which are associated with damage and plastic hardening, respectively, are considered in this work. For the sake of a schematic description of the above stated concepts, the uniaxial stress-strain curves shown in Figure 8.2 are used. In Figure 8.2, $\bar{E}$ is the initial undamaged Young's modulus, E is the damaged Young's modulus, S is the second Piola-Kirchhoff stress, and ε is the Lagrangian strain. Even though these notations are for the case of uniaxial state, they can be used in indicial tensor notation in the equations below without loss of generality. Referring to the solid curve in Figure 8.2, the total Lagrangian strain tensor ε is given by:

$$\varepsilon_{ij} = \varepsilon^p_{ij} + \varepsilon^e_{ij} + \varepsilon^d_{ij} \tag{8.116}$$

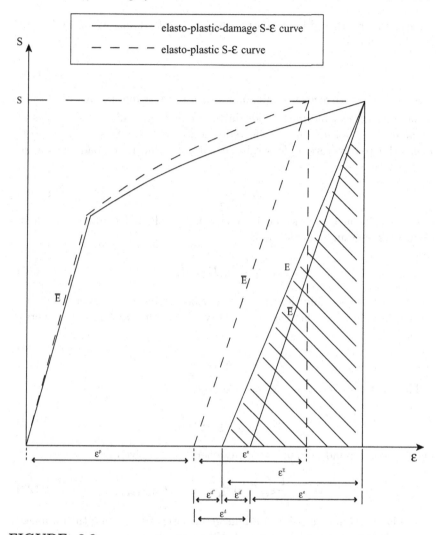

FIGURE 8.2

Schematic representation of elasto-plastic damage stress-strain curves for the uniaxial state of the stress.(Voyiadjis and Park, 1999)

where ε^p is the plastic strain tensor, ε^e is the elastic strain tensor, and ε^d is the additional strain tensor due to damage. Comparing equations (8.38) and (8.116), one notes that:

$$\varepsilon_{ij}^{dp} = \varepsilon_{ij}^{p} + \varepsilon_{ij}^{d} \tag{8.117}$$

Furthermore, the additional strain tensor due to damage can be decomposed as follows:

$$\varepsilon_{ij}^d = \varepsilon_{ij}^{d'} + \varepsilon_{ij}^{d''} \tag{8.118}$$

where $\varepsilon_{ij}^{d''}$ is the irrecoverable damage strain tensor due to lack of closure of the microcracks and microvoids during unloading, while $\varepsilon_{ij}^{d'}$ is the elastic damage strain tensor due to the reduction of the elastic stiffness tensor. Thus, the purely reversible strain tensor, ε^E, due to unloading can be obtained by:

$$\varepsilon_{ij}^E = \varepsilon_{ij}^e + \varepsilon_{ij}^{d'} \tag{8.119}$$

The strain energy ψ which is shown as the shaded triangular area in Figure 8.2 is assumed as follows:

$$\psi = \frac{1}{2\rho}\varepsilon_{ij}^E E_{ijkl}\varepsilon_{kl}^E \tag{8.120}$$

where ρ is the specific density. Furthermore, this strain energy can be decomposed into the elastic strain energy ψ^e and the damage strain energy ψ^d as follows:

$$\psi = \psi^e + \psi^d \tag{8.121}$$

The elastic strain energy ψ^e is given by:

$$\psi^e = \frac{1}{2\rho}\varepsilon_{ij}^e \bar{E}_{ijkl}\varepsilon_{kl}^e \tag{8.122}$$

and the corresponding damage strain energy ψ^d is given by:

$$\psi^d = \frac{1}{2\rho}\varepsilon_{ij}^E E_{ijkl}\varepsilon_{kl}^E - \frac{1}{2\rho}\varepsilon_{ij}^e \bar{E}_{ijkl}\varepsilon_{kl}^e \tag{8.123}$$

where $\bar{\mathbf{E}}$ and $\mathbf{E}$ are the initial undamaged elastic stiffness and the damaged elastic stiffness, respectively. These stiffnesses are defined such that:

$$\bar{E}_{ijkl} = \frac{\partial^2 \Psi}{\partial \varepsilon_{ij}^e \partial \varepsilon_{kl}^e} \tag{8.124}$$

$$E_{ijkl} = \frac{\partial^2 \Psi}{\partial \varepsilon_{ij}^E \partial \varepsilon_{kl}^E} \tag{8.125}$$

The damaged elastic stiffness in the case of finite deformation is given by Park and Voyiadjis (1998) as follows:

$$E_{ijrs} = N_{ikjl}\bar{E}_{klpq}N_{prqs} \tag{8.126}$$

where

$$N_{ikjl} = M_{ikjl}^{-1} \tag{8.127}$$
$$= a_{ik}^{-1} a_{jl}^{-1}$$

The elastic damage stiffness given by equation (8.126) is symmetric. This is in line with the classic sense of continuum mechanics which is violated by using the hypothesis of strain equivalence. Using the similar relation between the Lagrangian and the Eulerian strain tensors given by equation (8.42), the corresponding strain energy given by equation (8.120) can be written as follows:

$$\psi = \frac{1}{2\rho} \epsilon_{mn}^E F_{mi} F_{nj} E_{ijkl} F_{rk} F_{sl} \epsilon_{rs}^E \tag{8.128}$$
$$= \frac{1}{2\rho} \epsilon_{mn}^E \Lambda_{mnrs} \epsilon_{rs}^E$$

where ϵ^E is the Eulerian strain corresponding to the Lagrangian strain shown in equation (8.119), and Λ is termed the Eulerian elastic stiffness which is given by:

$$\Lambda_{mnrs} = F_{mi} F_{nj} E_{ijkl} F_{rk} F_{sl} \tag{8.129}$$

The second Piola-Kirchhoff stress tensor **S** is defined as follows:

$$S_{ij} = \rho \frac{\partial \psi}{\partial \varepsilon_{ij}^E} \tag{8.130}$$
$$= \rho \frac{\partial \psi^e}{\partial \varepsilon_{ij}^e}$$

The second Piola-Kirchhoff stress tensor **S** is related to the Cauchy stress tenor σ as follows:

$$S_{ij} = J F_{ik}^{-1} \sigma_{km} F_{jm} \tag{8.131}$$

The Kirchhoff stress tensor **T** is related to the Cauchy stress tensor σ as follows:

$$T_{ij} = J \sigma_{ij} \tag{8.132}$$

The rate of Helmholtz free energy is then given as follows:

$$d\Psi = d\psi + d\gamma \tag{8.133}$$

where $d\gamma$ is the rate of γ associated with the two neighboring constrained equilibrium states with two different sets of internal variables ϕ and $\mathbf{X}$. Using equation (8.120) or (8.121), the rate form of the strain energy can be given as follows since $d\bar{\mathbf{E}} = \mathbf{0}$:

$$\rho d\psi = \frac{1}{2} dE_{ijkl}\varepsilon_{ij}^{E}\varepsilon_{kl}^{E} + E_{ijkl}\varepsilon_{ij}^{E}d\varepsilon_{kl}^{E} - \frac{d\rho}{2\rho}\varepsilon_{ij}^{E}E_{ijkl}\varepsilon_{kl}^{E} \qquad (8.134)$$

or

$$\rho d\psi^{e} = d\varepsilon_{ij}^{e}\bar{E}_{ijkl}\varepsilon_{kl}^{e} - \frac{d\rho}{2\rho}\varepsilon_{ij}^{e}\bar{E}_{ijkl}\varepsilon_{kl}^{e} \qquad (8.135)$$

and

$$\rho d\psi^{d} = d\varepsilon_{ij}^{E}E_{ijkl}\varepsilon_{kl}^{E} + \frac{1}{2}\varepsilon_{ij}^{E}dE_{ijkl}\varepsilon_{kl}^{E} - d\varepsilon_{ij}^{e}\bar{E}_{ijkl}\varepsilon_{kl}^{e} - \frac{d\rho}{2\rho}(\varepsilon_{ij}^{E}E_{ijkl}\varepsilon_{kl}^{E} - \varepsilon_{ij}^{e}\bar{E}_{ijkl}\varepsilon_{kl}^{e}) \qquad (8.136)$$

If the deformation process is assumed to be isothermal with negligible temperature nonuniformities, the rate of the Helmholtz free energy can be written using the first law of thermodynamics (balance of energy) as follows:

$$d\Psi = T_{ij}D_{ij} - T\eta \qquad (8.137)$$

where T is the temperature and η is the irreversible entropy production rate. The product $T\eta$ represents the energy dissipation rate associated with both the damage and plastic deformation processes. The energy of the dissipation rate is given as follows:

$$T\eta = S_{ij}d\varepsilon_{ij}^{d''} + S_{ij}d\varepsilon_{ij}^{p} - d\gamma \qquad (8.138)$$

The first two terms on the right-hand side of the equation (8.138) represent a macroscopically nonrecoverable rate of work expanded on damage and plastic processes, respectively. Furthermore, the rate of the additional strain tensor due to damage is given by:

$$d\varepsilon_{ij}^{d} = d\varepsilon_{ij}^{d''} + d\varepsilon_{ij}^{d'} \qquad (8.139)$$

If we assume that the fraction of the additional strain tensor can be recovered during unloading, then the elastic damage tensor due to the reduction of the elastic stiffness is given by:

$$d\varepsilon_{ij}^{d'} = c\,d\varepsilon_{ij}^{d} \qquad (8.140)$$

where c is a fraction which ranges from 0 to 1. Then, the permanent damage strain due to lack of closure of microcracks and microcavities is given by:

$$d\varepsilon_{ij}^{d''} = (1 - c)d\varepsilon_{ij}^{d} \tag{8.141}$$

Thus, the energy of the dissipation rate given by equation (8.138) can be written as follows:

$$\begin{aligned} T\eta &= (1 - c)S_{ij}d\varepsilon_{ij}^{d} + S_{ij}d\varepsilon_{ij}^{p} - d\gamma \\ &= (1 - c)T_{ij}D_{ij}^{d} + T_{ij}D_{ij}^{p} - d\gamma \end{aligned} \tag{8.142}$$

The rate of energy associated with a specific microstructural change due to both the damage and the plastic processes can be decomposed as follows:

$$d\gamma = d\gamma^{d} + d\gamma^{p} \tag{8.143}$$

where one defines that

$$\rho d\gamma^{d} = Y_{ij}d\phi_{ij} \tag{8.144}$$

and

$$\rho d\gamma^{p} = A_{ij}dX_{ij} \tag{8.145}$$

where $\mathbf{Y}$ and $\mathbf{A}$ are the general forces conjugated by damage and plastic yielding, respectively. They are defined as:

$$\begin{aligned} Y_{ij} &= \rho\frac{\partial\Psi}{\partial\phi_{ij}} \\ &= \rho\frac{\partial\psi^{ed}}{\partial\phi_{ij}} + \rho\frac{\partial\gamma}{\partial\phi_{ij}} \end{aligned} \tag{8.146}$$

$$A_{ij} = \rho\frac{\partial\Psi}{\partial X_{ij}} \tag{8.147}$$

In view of equation (8.142), one notes that it is equivalent to the work by Lubarda and Krajcinvic (1995) when $(1 - c) = \frac{1}{2}$. A schematic representation of the elastic, damage, and plastic strain rates, and the total rate of work $S_{ij}d\epsilon_{ij}$ is shown on the uniaxial stress-strain curve in Figure 8.3.

8.6 Constitutive Equation for Finite Elasto-Plastic Deformation with Damage Behavior

The kinematics and the thermodynamics discussed in the previous sections provide the basis for a finite deformation elasto-plasticity. In this section, the basic structure of the constitutive equations is reviewed based on the generalized Hooke's law, originally obtained for small elastic strains such that the second Piola-Kirchhoff stress tensor $\mathbf{S}$ is the gradient of free energy Ψ with respect to the Lagrangian elastic strain tensor ε^E given by equation (8.130). Referring to Figure 8.2, one obtains the following relation when generalized to the three dimensional state of stress and strain:

$$S_{ij} = \bar{E}_{ijkl}(\varepsilon_{kl} - \varepsilon_{kl}^p - \varepsilon_{kl}^d) \qquad (8.148a)$$

$$= \bar{E}_{ijkl}\varepsilon_{kl}^e \qquad (8.148b)$$

$$= E_{ijkl}(\varepsilon_{kl}^e + \varepsilon_{kl}^{d'}) \qquad (8.148c)$$

$$= E_{ijkl}(\varepsilon_{kl} - \varepsilon_{kl}^{d''} - \varepsilon_{kl}^p) \qquad (8.148d)$$

From the incremental analysis, one obtains the following rate form of the constitutive equation by differentiating equation (8.148a):

$$dS_{ij} = \bar{E}_{ijkl}(d\varepsilon_{kl} - d\varepsilon_{kl}^p - d\varepsilon_{kl}^d) \qquad (8.149)$$

Consequently the constitutive equation of the elasto-plastic damage behavior can be writen as follows:

$$dS_{ij} = E_{ijkl}^{Dp}d\varepsilon_{kl} \qquad (8.150)$$

where $\mathbf{E}^{Dp}$ is the damage elasto-plastic stiffness tensor and is expressed as follows:

$$E_{ijkl}^{Dp} = \bar{E}_{ikjl} - E_{ikjl}^p - E_{ikjl}^d \qquad (8.151)$$

where $\mathbf{E}^p$ is the plastic stiffness and $\mathbf{E}^d$ is the damage stiffness. Both $\mathbf{E}^p$ and $\mathbf{E}^d$ are the reduction in stiffness due to the plastic and damage deteriorations, respectively. The plastic stiffness and the damage stiffness can be obtained by using the flow rule and damage evolution law, respectively. By assuming that the reference state coincides with the current configuration, the second Piola-Kirchhoff stress rate $d\mathbf{S}$ can be replaced by the corotational rate of the Cauchy stress tensor $\boldsymbol{\sigma}$ and the rate of the Lagrangian strain tensor $d\boldsymbol{\epsilon}$ by the deformation rate $\mathbf{D}$ as follows:

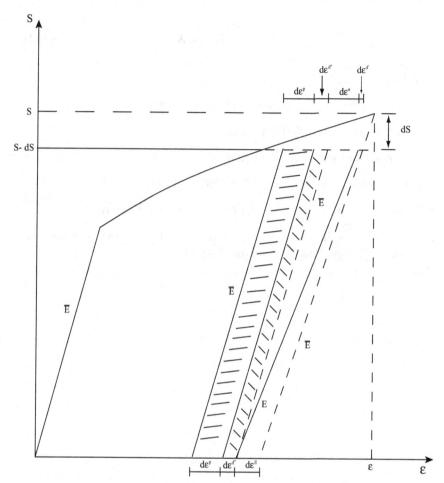

FIGURE 8.3
Schematic representation of elasto-plastic damage strain increments in the
case of a uniaxial stress-strain curve.(Voyiadjis and Park, 1999)

$$\overset{\circ}{\sigma}_{ij} = E^{Dp}_{ijkl} D_{kl} \qquad (8.152)$$

The corotational rate of the Cauchy stress tensor $\boldsymbol{\sigma}$ is related to the rate
of the Cauchy stress tensor $d\boldsymbol{\sigma}$ as follows:

$$\overset{\circ}{\sigma}_{ij} = d\sigma_{ij} - W^*_{ik}\sigma_{kj} + \sigma_{ik}W^*_{kj} \qquad (8.153)$$

where

$$\mathbf{W}^* = \mathbf{W} - \mathbf{W}^{\mathbf{p}} - \mathbf{W}^{\mathbf{d}} \qquad\qquad (8.154)$$

Problems

8.1. Derive the matrix in equation (8.19) explicitly.

8.2. Derive equation (8.55) explicitly.

8.3. Derive equation (8.70) explicitly.

8.4. Derive equations (8.128) and (8.129) explicitly.

8.5. Derive equations (8.134), (8.135), and (8.136) explicitly.

8.6. Derive an explicit formula for the damage elasto-plastic stiffness tensor E_{ijkl}^{Dp} appearing in the equation (8.150).

References

Allen, D. H., Harris, C. E. and Groves, S. E. (1987) A Thermomechanical Constitutive Theory for Elastic Composites with Distributed Damage-I. Theoretical Development, International Journal of Solids and Structures, Vol. 23, No. 9, pp. 1301-1318.

Armstrong, P. J. and Frederick, C. O. (1966) A mathematical representation of the multiaxial Bauschinger effect, CEGB Report RD/B/N/731, Berkeley Nuclear Laboratories.

Asaro, R. (1983) Micromechanics of Crystals and Polycrystals, Advances in Applied Mechanics, Vol. 23, pp. 1-115.

Bammann, D. and Aifantis, E. (1989) A Damage Model for Ductile Metals, Nuclear Engineering and Design, Vol. 116, pp. 13-32.

Barsoum, R. S. (1976) On the Use of Isoparametric Finite Elements in Linear Fracture Mechanics, International Journal of Numerical Method Engineering, Vol. 10, pp. 25 37.

Betten, J. (1981) Damage Tensors in Continuum Mechanics, J. Mecanique Theorique et Appliquees, Vol. 2, pp. 13-32 (Presented at Euromech Colloqium 147 on Damage Mechanics, Paris-VI, Cachan, 22 September).

Betten, J. (1982) Representation of Constitutive Equations in Creep Mechanics of Isotropic and Anisotropic Materials, Presentation at IUTAM, 3rd Symposium on Creep in Structures, Leicester, September 8-12, 1980, published in Proceedings, Creep in Structures, edited by A. R. S. Pointer, Springer-Verlag, Berlin, pp. 179-201.

Betten, J. (1983) Damage Tensors in Continuum Mechanics, Journal de Machanique Theorique et Appliquee, Vol 2, No. 1, pp. 13-32.

Betten, J. (1986) Applications of Tensor Functions to the Formulation of Constitutive Equations Involving Damage and Initial Anisotropy, Engineering Fracture Mechanics, Vol. 25, Nos. 5/6, pp. 573-584.

Boyd, J. G., Costanzo, F. and Allen, D. H. (1983) A Micromechanics Approach for Constructing Locally Averaged Damage Dependent Constitutive Equations in Inelastic Composites, International Journal of Damage Mechanics, Vol. 2, pp. 209-228.

Chaboche, J. L. (1974) Une Loi Differentielle d'Endommagement de Fatigue avec Cumulation Nonlineare, Rev. Francais Mecanique, No. 50-51 (in French).

Chaboche, J. L. (1977) Sur l'utilisation des variables d'etat interne pour la description du comportement viscoplastique et de la rupture par endommagement, In Symposium Franco-polonais de Rheologie et Mecanique, Cracovie.

Chaboche, J. L. (1979) Le concept de contrainte effective applique a l' elasticite et a la viscoplasticite en presence d'un endommagement anisotrope, In Colooque Euromech 115, Villard de Lans.

Chaboche, J. (1981) Continuous Damage Mechanics: A Tool to Describe Phenomena Before Crack Initiation, Nuclear Engineering and Design, Vol. 64, pp. 233-247.

Chaboche, J. (1988a) Continuum Damage Mechanics: Part I - General Concepts, Journal of Applied Mechanics, ASME, Vol. 55, pp. 59-64.

Chaboche, J. (1988b) Continuum Damage Mechanics: Part II - Damage Growth, Crack Initiation, and Crack Growth, Journal of Applied Mechanics, ASME, Vol. 55, pp. 65-72.

Chaboche, J. and Lesne, P. (1988) A Nonlinear Continuous Fatigue Damage Model, Fatigue and Fracture of Engineering Materials and Structures, Vol. 11, No. 1, pp. 1-17.

Chaboche, J. L. (1989) Constitutive equations for cyclic plasticity and cyclic viscoplasticity, International Journal of Plasticity Vol. 5, pp. 247-302.

Chaboche, J. L. (1991) On some modifications of kinematic hardening to improve the description of ratchetting effects, International Journal of Plasticity Vol. 7, pp. 661-678.

Chaboche, J. (1993) Cyclic Viscoplastic Equations: Part I - Thermodynamically Consistent Formulation, ASME Journal of Applied Mechanics, Vol. 60, pp. 813-821.

Chow, C. and Lu, T. (1989) On Evolution Laws of Anisotropic Damage, Engineering Fracture Mechanics, Vol. 34, No. 3, pp. 679-701.

Chow, C. and Wang, J. (1987) An Anisotropic Theory of Continuum Damage Mechanics for Ductile Fracture, Engineering Fracture Mechanics, Vol. 27, pp. 547-558.

Chow, C. and Wang, J. (1987) An Anisotropic Theory of Elasticity for Continuum Damage Mechanics, International Journal of Fracture, Vol. 33, pp. 3-16.

Chow, C. and Wang, J. (1988a) Ductile Fracture Characterization with an Anisotropic Continuum Damage Theory, Engineering Fracture Mechanics, Vol. 30, pp. 547-563.

Chow, C. and Wang, J. (1988b) A Finite Element Analysis of Continuum Damage Mechanics for Ductile Fracture, International Journal of Fracture, Vol. 38, pp. 83-102.

Chow, C. L. and Wei, Y. (1991) A Damage Mechanics Model of Fatigue Crack Initiation in Notched Plates, Theoretical and Applied Fracture Mechanics, Vol. 16, pp. 123-133.

Chrzanowski, M. (1973) The Description of Metallic Creep in the Light of Damage Hypothesis and Strain Hardening, Diss. hab. Politechnika Krakowska.

Chung, T. (1996) Applied Continuum Mechanics, Cambridge University Press, United Kingdom.

Cordebois, J. P. (1983) Criteres d' Instabilite Plastiques et Endommagement Ductile en Grandes Deformations, These de Doctorat, Presente a l'Universite Pierre et Marie Curie.

Cordebois, J. and Sidoroff, F. (1979) Damage Induced Elastic Anisotropy, Colloque Euromech, 115, Villard de Lans.

Cordebois, J. P. and Sidoroff, F. (1979) Damage Induced Elastic Anisotropy, In J. P. Boehler, editor, Mechanical Behavior of Anisotropic Solids, Colloque Euromech 115, Villard-deLans, June 19-22, pp. 761-774, Martinus Nijhoff Publishers.

Cordebois, J. and Sidoroff, F. (1982) Anisotropic Damage in Elasticity and Plasticity, J. Mech. Theor. Appl. (Special Edition), pp. 45-60 (in French).

Corless, R. (1995) Essential Maple: An Introduction for Scientific Programmers. Springer-Verlag, Germany.

Cornil, J. and Testud, P. (2001) An Introduction to Maple V, Springer-Verlag, Germany.

Dafalias, Y. F. (1981) The concept and application of the bounding surface in plasticity theory, Physical Nonlinearities in Structural Analysis, (Hult, J., Lamaitre, J., ed.), IUTAM Symposium, Senlis, France Springer Verlag, pp. 56-63.

Dafalias, Y. (1983) Corotational Rates for Kinematic Hardening at Large Plastic Deformations, Journal of Applied Mechanics, ASME, Vol. 50, pp. 561-565.

Doghri, I. (1988) Fully Implicit Integration and Consistent Tangent Modules in Elasto-Plasticity, Mechanics of Solid Materials, Vol. 55, pp. 59-64.

Drucker, H. (1948) A More Fundamental Approach to Plastic Stress-Strain Relations, Proceedings of the First U.S. National Congress in Applied Mechanics, ASME, New York, pp. 487-491.

Drucker, H. (1959) A Definition of Stable Inelastic Material, Journal of Applied Mechanics, Transactions ASME, Vol. 26, p 101.

Eisenburg, M. and Chen, C. (1984) The Anisotropic Deformation of Yield Surfaces, ASME Journal of Engineering Material Technology, Vol. 106, pp. 355-360.

Goel, R. P. (1975) On the Creep Rupture of a Tube and a Sphere, J. Applied Mechanics, September, Transactions of the ASME, pp. 625-629.

Hayhurst, D. R., Trampczynski, W. A. and Leckie, F. A. (1980) Creep Rupture under Non-Proportional Loading, Acta Metallurgica, Vol. 28, pp. 1171-1183.

Eringen, A. (1980) Mechanics of Continua, Krieger, New York, USA.

Freed, A., Chaboche, J., and Walker, K. (1991) A Viscoplastic Theory of with Thermodynamic Consideration, Acta Mechanica, Vol. 90, pp. 219-241.

Fung, Y. (1965) Foundations of Solid Mechanics, Prentice Hall, New Jersey, USA.

Fung, Y. (1994) A First Course in Continuum Mechanics, Third Edition, Prentice Hall, New Jersey, USA.

Green, A. and Zerna, W. (1968) Theoretical Elasticity, Second Edition, Dover, New York, USA.

Gurson, A. (1977) Continuum Theory of Ductile Rupture by Void Nucleation and Growth - Part I: Yield Criteria and Flow Rules for Porous Ductile Media, Journal of Engineering Materials and Technology, Vol. 99, No. 2, pp. 2-15.

Hayhurst, D. (1972) Creep Rupture Under Multiaxial States of Stress, Journal of the Mechanical and Physics of Solids, Vol. 20, pp. 381-390.

Heck, A. (1996) Introduction to Maple, Second Edition, Springer-Verlag, Germany.

Henshell, R. D. and Shaw, K. G. (1975) Crack Tip Finite Elements are Unnecessary, International Journal of Numerical Method Engineering, Vol. 9, pp. 495-507.

Hjelmstad, K. (1997) Fundamentals of Structural Mechanics, Prentice Hall, New Jersey, USA.

Hori, H. and Nemat-Nasser, S. (1983) Overall Moduli of Solids with Microcracks: Load-Induced Anisotropy, Journal of Mechanics and Physics of Solids, Vol. 31, pp. 155-171.

Hult, J. (1974) Creep in Continua and Structures, in Topics in Applied Continuum Mechanics (Edited by Zeman and Ziegler), pp. 137, Springer, N.Y.

Hult, J. (1979) CDM: Capabilities, Limitations, and Promises in Mechanisms of Deformation and Fracture, Edited by K. Easterling, Pergamon Press, Oxford, pp. 233-247.

Ilyushin, A. A. (1954) on relationship between stress and small strain in continuum mechanics, (in, Russian) Prik. Mate. Mech. Vol. 18, pp. 641-666.

Johnson, A. E. (1960) Complex-Stress Creep of Metals, Metallurgical Reviews, Vol. 5, No. 20, pp. 447-506.

Johnson, G. and Bammann, D. (1984) A Discussion of Stress Rates in Finite Deformation Problems, International Journal of Solids and Structures, Vol. 20, No. 8, pp. 725-737.

Ju, J. (1989) Energy Based Coupled Elasto-plastic Damage Theories: Constitutive Modeling and Computational Aspects, International Journal of Solids and Structures, Vol. 25, pp. 803-833.

Ju, J. W. (1989) Energy-Based Coupled Elastoplasic Damage Models at Finite Strains, Journal of Engineering Mechanics, Vol. 115, No. 11, pp. 2507-2525.

Ju, J. W. (1990) Isotropic and Anisotropic Damage Variables in Continuum Damage Mechanics, Journal of Engineering Mechanics, Vol. 116, No. 12, pp. 2764-2770.

Ju, J. W. and Lee, X. (1991) Micromechanical Damage Models for Brittle Solids, I: Tensile Loadings, Journal of Engineering Mechanics, Vol. 117, No. 7, pp. 1495-1514.

Ju, J. W. and Chen, Tsung-Muh (1994) Effective Elastic Moduli of Two-Dimensional Brittle Solids with Interacting Microcracks, Part I: Basic Formulations, Journal of Applied Mechanics, Vol. 61, pp. 349-357.

Ju, J. W. and Chen, Tsung-Muh (1994) Effective Elastic Moduli of Two-Dimensional Brittle Solids with Interacting Microcracks, Part I: Evolutionary Damage Models, Journal of Applied Mechanics, Vol. 61, pp. 358-366.

Kachanov, L. M. (1958) On the Creep Fracture Time, Izv Akad, Nauk USSR Otd, Tekh., Vol. 8, pp. 26-31 (in Russian).

Kachanov, L. M. (1986) Introduction to Continuum Damage Mechanics, Martinus Nijhoff Publishers, The Netherlands.

Kattan, P. I. and Voyiadjis, G. Z. (1990) A Coupled Theory of Damage Mechanics and Finite Strain Elasto-Plasticity, Part I: Damage and Elastic

Deformations, International Journal of Engineering Science, Vol. 28, No.5, pp. 421-435.

Kattan, P. and Voyiadjis, G. (1993) A Plasticity-Damage Theory for Large Deformation of Solids - Part II: Applications to Finite Simple Shear, International Journal of Engineering Science, Vol. 31, No. 1, pp. 183-199.

Kattan, P. and Voyiadjis, G. (2001a) Damage Mechanics with Finite Elements: Practical Applications with Computer Tools, Springer-Verlag, Germany.

Kattan, P. and Voyiadjis, G. (2001b) Decomposition of Damage Tensor in Continuum Damage Mechanics, ASCE Journal of Engineering Mechanics, Vol. 127, No. 9, pp. 940-944.

Kofler, M. (1997) Maple: An Introduction and Reference, Addison-Wesley, United Kingdom.

Krajcinovic, D. (1983) Constitutive Equations for Damaging Materials, Journal of Applied Mechanics, ASME, Vol. 50, pp. 355-360.

Krajcinovic, D. (1984) Continuum Damage Mechanics, Applied Mechanics, Reviews, Vol. 37, pp. 1-6.

Krajcinovic, D. (1985) Continuous Damage Mechanics Revisited: Basic Concepts and Definitions, Journal of Applied Mechanics, Vol. 52, pp. 829-834.

Krajcinovic, D. and Foneska, G. U. (1981) The Continuum Damage Theory for Brittle Materials, Journal of Applied Mechanics, ASME, Vol. 48, pp. 809-824.

Krajcinovic, D. and Mastilovic, S. (1995) Some Fundamental Issues of Damage Mechanics, Mechanics of Materials, Vol. 21, pp. 217-230.

Kyoya, T., Ichikawa, Y, and Kawamoto, T. (1985) A Damage Mechanics Theory for Discontinuous Rock Mass, Numerial Methods in Geomechanics, Roterdam, pp. 469-480.

Ladeveze, P. and Lemaitre, J. (1984) Damage Effective Stress in Quasi-Unilateral Conditions, The 16th International Congress of Theoretical and Applied Mechanics, Lyngby, Denmark.

Lai, W., Rubin, D., and Krempl, E. (1984) Introduction to Continuum Mechanics, Revised Edition, Pergamon Press, United Kingdom.

Leckie, F. A. and Hayhurst, D. (1974) Creep Rupture of Stresses, Proceedings of the Royal Society, London, Vol. A340, pp. 323-347.

Leckie, F. A. and Hayhurst, D. R. (1975) The Damage Concept in Creep Mechanics, Mechanics Research Communications, Vol. 2, pp. 23-26.

Leckie, F. and Onat, E. (1980) Tensorial Nature of Damage Measuring Internal Variables, in IUTAM Colloquim on Phyisical Nonlinearities in Structural Analysis, pp. 140-155, Springer-Verlag, Germany.

Leckie, F. A. and Ponter, A. R. S. (1974) On the State Variable Description of Creeping Materials, Ingenieur Archiv, Vol. 43, pp. 158-167.

Leckie, F. A. (1993) Tensorial Nature of Damage Measuring Internal Variables, International Journal of Solids and Structures, Vol. 30, No. 1, pp. 19-36.

Lee, E. (1981) Some Comments on Elastic-Plastic Analysis, International Journal of Solids and Structures, Vol. 17, pp. 859-872.

Lee, H., Li, G., and Lee, S. (1986) The Influence of Anisotropic Damage on the Elastic Behavior of Materials, International Seminar on Local Approach of Fracture, Moret-sur-Loing, France, pp. 79-90.

Lee, E., Mallet, R., and Wertheimer, T. (1983) Stress Analysis for Anisotropic Hardening in Finite-Deformation Plasticity, Journal of Applied Mechanics, ASME, Vol. 50, pp. 554-560.

Lee, H., Li, G. and Lee, S. (1986) The Influence of Anisotropic Damage on the Elastic Behavior of Materials, In International Seminar on Local Approach of Fracture, Moret-sur-Laing, France, pp. 79-90.

Lee, H., Peng, K., and Wang, J. (1985) An Anisotropic Damage Criterion for Deformation Instability and its Application to Forming Limit Analysis of Metal Plates, Engineering Fracture Mechanics, Vol. 21, pp. 1031-1054.

Lee, X. and Ju, J. W. (1991) Micromechanical Damage Models for Brittle Solids, II: Compressive Loadings, Journal of Engineering Mechanics, Vol. 117, No. 7, pp. 1515-1536.

Lemaitre, J. (1971) Evaluation of Dissipation and Damage in Metals Subjected to Dynamic Loading, Proceedings of I.C.M., Kyoto, Japan.

Lemaitre, J. (1984) How to Use Damage Mechanics, Nuclear Engineering and Design, Vol. 80, pp. 233-245.

Lemaitre, J. (1985) A Continuous Damage Mechanics Model of Ductile Fracture, Journal of Engineering Materials and Technology, Vol. 107, No. 42, pp. 83-89.

Lemaitre, J. (1986) Local Approach to Fracture, Engineering Fracture Mechanics, Vol. 25, Nos. 5/6, pp. 523-537.

Lemaitre, J. (1996) A Course on Damage Mechanics, Second Edition, Springer-Verlag, Germany.

Lemaitre, J. and Chaboche, J. L. (1974) A Nonlinear Model of Creep Fatigue Cumulation and Interaction, Proc. IUTAM, Symposium on Mechanics of Viscoelastic Media Bodies, Edited by Hult, pp. 291-301.

Lemaitre, J. and Chaboche, J. L. (1975) A Nonlinear Model of Creep Fatigue Cumulation and Interaction, Proc. IUTAM, Symposium on Mechanics of Viscoelastic Media and Bodies (Edited by Hult), pp. 291-301.

Lemaitre, J. and Chaboche, J. L. (1978) Aspect Phenomenologique de la Rupture par Endommagement, Journal de Mecanique Appiquee, Vol. 2, pp. 317-365.

Lemaitre, J. and Chaboche, J. L. (1985) Mecanique de Materiaux Solides, Dunod, Paris.

Lemaitre, J. and Chaboche, J. (1994) Mechanics of Solid Materials, Cambridge University Press, United Kingdom.

Lemaitre, J. and Dufailly, J. (1987) Damage Measurements, Engineering Fracture Mechanics, Vol. 28, Nos. 5/6, pp. 643-661.

Loret, B. (1983) On the Effects of Plastic Rotation in the Deformation of Anisotropic Elasto-Plastic Materials, Mechanics of Materials Journal, Vol. 2, pp. 287-304.

Lu, T. J. and Chow, C. L. (1990) On Constitutive Equations of Inelastic Solids with Anisotropic Damage, Theoretical and Applied Fracture Mechanics, Vol. 14, pp. 187-218.

Lubarda, V. and Krajcinovic, D. (1995) Some Fundamental Issues in Rate Theory of Damage Elasto-plasticity, International Journal of Plasticity, Vol. 11, No. 7, pp. 763-797.

Lubarda, V. A., Krajcinovic, D. and Mastilovic, S. (1994) Damage Model for Brittle Elastic Solids with Unequal Tensile and Compressive Strengths, Engineering Fracture Mechanics, Vol. 49, No. 5, pp. 681-697.

Lubliner, J. (1990) Plasticity Theory, Macmillan, New York, USA.

Mandel, J. (1973) Relations de Comportment des Millieux Elastiques - Plastiques et Elastiques - Viscoplastiques. Notion de Repere Directeur Foundations of Plasticity, A. Sawczu, (editor), Noordhoff, Leyden, pp. 387-400.

Marsden, J. and Hughes, T. (1983) Mathematical Foundations of Elasticity, Dover, New York, USA.

Maugin, G. (1992) The Thermomechanics of Plasticity and Fracture, Cambridge University Press, United Kingdom.

McDonald, P. (1996) Continuum Mechanics, PWS, ITP, Boston, USA.

McDowell, D. L. (1987) An evaluation of recent developments in hardening and flow rules for rate independent cyclic plasticity, J. Appl. Mech Vol. 54, pp. 323-334.

Mroz, Z. (1967) On the description of anisotropic workhardening, Journal of Mechanical Physics and Solids Vol. 15, pp. 163.

Mroz, Z. (1969) An attempt to describe the behavior of metals under cyclic loads using a more general workhardening model, Acta Mechanica Vol. 7, pp. 199-212.

Mroz, Z. (1973) <u>Mathematical Models of Inelastic Material Behavior</u>, University of Waterloo, pp. 120-146.

Murakami, S. (1983) Notion of Continuum Damage Mechanics and its Application to Anisotropic Creep Damage Theory, Journal of Engineering Materials and Technology, Vol. 105, pp. 99-105.

Murakami, S. (1988) Mechanical Modeling of Material Damage, Journal of Applied Mechanics, ASME, Vol. 55, pp. 280-286.

Murakami, S. and Imaizumi, T. (1982) Mechanical Description of Creep Damage State and its Experimental Verification, Journal de Mecanique Theorique et Appliquee, Vol. 1, pp. 743-761.

Murakami, S. and Ohno, N. (1978) Creep Damage Analysis in Thin-Walled Tubes, in Inelastic Behavior of Pressure Vessel and Piping Components, edited by T. Y. Chang and E. Krempl, PVP-PB-028, ASME, New York, pp. 55-69.

Murakami, S. and Ohno, N. (1980) A Continuum Theory of Creep and Creep Damage, In A. R. S. Ponter and D. R. Hayhurst, editors, Creep in Structures, IUTAM 8rd Symposium, Leicester, UK, September 8-12, pp. 422-443. Springer Verlag.

Murkami, S. and Ohno, N. (1981) A Continuum Theory of Creep and Creep Damage, in Proceedings of the Third IUTAM Symposium on Creep in Structures, pp. 422-444, Springer-Verlag, Germany.

Murakami, S., Sanomura, Y. and Saitoh, K. (1986) Formulation of Cross-Hardening in Creep and its Effect on the Creep Damage Process of Copper, ASME Journal of Engineering Materials and Technology, Vol. 108, pp. 167-173.

Nermat-Nasser, S. (1979) Decomposition of Strain Measures and Their Rates in Finite Deformation Elastopasticity, International Journal of Solids and Structures, Vol. 15, pp. 155-166.

Nermat-Nasser, S. (1983) On Finite Plastic Flow of Crystalline Solids and Geomaterials, Journal of Applied Mechanics, ASME, Vol. 50, pp. 1114-1126.

Nicolaides, R. and Walkington, N. (1996) <u>Maple: A Comprehensive Introduction</u>, Cambridge University Press, United Kingdom.

Ohno, N. (1982) A constitutive model of cyclic plasticity with a non-hardening strain region, ASME Journal of Applied Mechanics Vol. 49, pp. 721-727.

Ohno, N. and Wang, J. D. (1991) Two equivalent forms of nonlinear kinematic hardening: application to nonisothermal plasticity, International Journal of Plasticity Vol. 7, pp. 637-650.

Ohno, N. and Wang, J. D. (1993) Kinematic hardening rules with critical state of dynamic recovery, Part I: Formulation and basic features for ratchetting behavior, International Journal of Plasticity Vol. 9, No. 375-390.

Oldroyd, J. (1950) On the Foundation of Rheological Equations of State, Proceedings of the Royal Society, London, Vol. A200, pp. 523-541.

Onat, E. T. (1986) Representation of Mechanical Behavior in the Presence of Internal Damage, Engineering fracture Mechanics, Vol. 25, No. 5/6, pp. 605-614.

Onat, E. and Leckie, F. (1988) Representation of Mechanical Behavior in the Presence of Changing Internal Structure, Journal of Applied Mechanics, ASME, Vol. 55, pp. 1-10.

Ortiz, M. and Simo, C. (1986) An Analysis of a New Class of Integration Algorithms for Elasto-plastic Constitutive Relations, International Journal for Numerical Methods in Engineering, Vol. 23, pp. 353-366.

Park, T. and Voyiadjis, G. Z., (1998) "Kinematic Description of Damage", Journal of Applied Mechanics, ASME, Vol. 65, pp. 93-98.

Paulun, J. and Pecherski, R. (1985) Study of Corotational Rates for Kinematic Hardening of in Finite-Deformation Plasticity, Archives of Mechanics, Vol. 37, No. 6, pp. 661-677.

Perzyna, P. (1963) The Constitutive Equations for Rate Sensitive Plastic Material, Applied Mathematics, Vol. 20, pp. 321-332.

Perzyna, P. (1971) Thermodynamic Theory of Viscoplasticity, Advances in the Applied Mathematics, Vol. 11, pp. 313-345.

Phillips, A., Tang, J. L. and Ricciuti, M. (1974) Some new observations on yield surfaces, Acta Mechanica Vol. 20, pp. 23-39.

Rabotnov, Y. N. (1968) Creep Rupture, in Proceedings, Applied Mechanics Conference, Stanford University, edited by M. Hetenyi and H. Vincenti, pp. 342-349.

Rabotnov, Y. N. (1969) Creep Problems of Structural Mechanics, North Holland, Amsterdam.

Rivlin, R. S. and Smith, G. F. (1969) Orthogonal Integrity Basis for N Symmetric Matrices, in Contributions to Mechanics, edited by D. Abir, Pergamon Press, Oxford, pp. 121-141.

Rubin, M. (1982) A Thermoelastic Viscoplastic Model with a Rate Dependent Yield Strength, ASME Journal of Applied Mechanics, Vol. 49, pp. 305-311.

Schwartz, D. (1999) An Introduction to Maple, Prentice Hall, New Jersey, USA.

Segel, L. (1987) Mathematics Applied to Continuum Mechanics, Dover, New York, USA.

Shiratori, E., Ikegami, K. and Yoshida, F. (1979) Analysis of stress-strain relations of use of an anisotropic hardening potential, J. Mech. Physics Solids Vol. 27, pp. 213-229.

Sidoroff, F. (1979) Description of Anisotropic Damage Application to Elasticity, In JeanPaul Boehler, editor, Mechanical Behavior of Anisotropic Solids / N295 Comportement Mecanique Des Solides Anisotropes. Martinus Nijhoff Publishers, 1979. Proceedings of the Euromech Colloquim 115, N295 Villard-de-Lans, June 19-22, France.

Sidoroff, F. (1980) Description of Anisotropic Damage Application to Elasticity, In J. Hult and J. Lemaitre, editors, Physical Non-Linearities in Structural Analysis, IUTAM Series, pp. 237-244, Springer-Verlag.

Sidoroff, F. (1981) Description of Anisotropic Damage Application to Elasticity, in IUTAM Colloqium on Physical Nonlinearities in Structural Analysis, pp. 237-244, Springer-Verlag, Germany.

Simo, J. and Ju, J. (1987) Strain and Stress-Based Continuum Damage Models - Part I: Formulation, International Journal of Solids and Structures, Vol. 23, No. 7, pp. 821-840.

Simo, J. and Pister, K. (1984) Remarks on Rate Constitutive Equations for Finite Deformation Problems: Computational Implications, Computer Methods in Applied Mechanics and Engineering, Vol. 46, pp. 201-215.

Smith, G. F. (1971) On Isotropic Functions of Symmetric Tensors, Skew-Symmetric Tensors, and Vectors, International Journal of Engineering Science, Vol. 9, pp. 899-916.

Spencer, A. J. M. (1971) Theory of Invariants, in Continuum Physics, edited by A. C. Eringen, Vol. 1, Mathematics, Academic Press, New York and London.

Spencer, A. J. M. and Rivlin, R. S. (1958/1959a) The Theory of Matrix Polynomials and its Application to the Mechanics of Isotropic Continua, Arch. Rational Mech. Anal., Vol. 2, pp. 309-336.

Spencer, A. J. M. and Rivlin, R. S. (1958/1959b) Finite Integrity Bases for Five or Fewer Symmetric 3 x 3 Matrices, Arch. Rational Mech. Anal., Vol. 2, pp. 435-446.

Srivatsavan, P. and Subramanyan, S. (1978) A Cumulative Damage Rule Based on Successive Reduction in Fatigue Limit, Journal of Engineering Mechanics and Technology, Vol. 100, pp. 212-214.

Stolz, C. (1986) General Relationships between Micro and Macro Scales for the Nonlinear Behavior of Heterogeneous Media, in Modeling Small Deformations of Polycrystals, Edited by J. Gittus and J. Zarka, pp. 89-115, Elsevier, The Netherlands.

Stumvoll and Swoboda (1993) Deformation Behavior of Ductile Solids Containing Anisotropic Damage, ASCE Journal of Engineering Mechanics, Vol. 119, No. 7, pp. 169-192.

Subramanyan, S. (1976) A Cumulative Damage Rule Based on the Knee Point of the S-N Curve, Journal of Engineering Mechanics and Technology, Vol. 98, pp. 316-321.

Tsamasphyras, G. and Giannakopoulos, A. E. (1989) The Optimum Finite Element Grids Around Crack Singularities in Bilinear Elasto-plastic Materials, Engineering Fracture Mechanics, Vol. 32, No.4, pp. 515-522.

Tseng, N. T. and Lee, G. C. (1983) Simple plasticity model of the two-surface type, ASCE J. Engineering Mechanics Vol. 109, No. 3, pp. 795-810.

Tvergaard, V. (1982) Material Failure by Void Coalescence in Localized Shear Bands, International Journal of Solids and Structures, Vol. 18, pp. 659-672.

Tvergaard, V. and Needleman, A. (1984) Analysis of Cup-Cone Fracture in a Round Tensile Bar, Acta Metallurgica, Vol. 32, p. 157.

Voyiadjis, G. Z. (1984) Experimental Determination of the Material Parameters of the Elasto-Plastic Work Hardening of Metal Alloys, Material Science and Engineering Journal, Vol. 62, No. 1, pp. 99-107.

Voyiadjis, G. Z. (1988) Degradation of Elastic Modulus in Elasto-Plastic Coupling with Finite Strains, International Journal of Plasticity, Vol. 4, pp. 335-353.

Voyiadjis, G. Z. and Basuroychowdury, I. N. (1998) A Plasticity Model for Multiaxial Cyclic Loading and Ratcheting, Acta Mechanica, Vol. 126, pp. 19-35.

Voyiadjis, G. Z. and Foroozesh, M. (1990) An Anisotropic Distortional Yield Model, ASME Journal of Applied Mechanics, Vol. 57, pp. 537-547.

Voyiadjis, G. Z. and Guelzim, Z. (1996) A Coupled Incremental Damage and Plasticity Theory for Metal Matrix Composites, Journal of the Mechanical Behavior of Materials, Vol. 6, No. 3, pp. 193-219.

Voyiadjis, G. Z. and Kattan, P. I. (1989) Eulerian Constitutive Model for Finite-Strain Plasticity with Anisotropic Hardening, Mechanics of Material Journal, Vol. 7, No. 4, pp. 279-293.

Voyiadjis, G. Z. and Kattan, P. I.(1990) A Coupled Theory of Damage Mechanics and Finite-Strain Elasto-Plasticity - Part II: Damage and Finite-Strain Plasticity, International Journal of Engineering Science, Vol. 28, No. 6, pp. 505-524.

Voyiadjis, G. Z. and Kattan, P. I. (1992) A Plasticity-Damage Theory for Large Deformation of Solids - Part I: Theoretical Formulation, International Journal of Engineering Science, Vol. 30, No. 9, pp. 1089-1108.

Voyiadjis, G. Z. and Kattan, P. I. (1993) Anisotropic Damage Mechanics Modeling in Metal Matrix Composites, Technical Report, Final Report Submitted to the Air Force Office of Scientific Research.

Voyiadjis, G. Z. and Kattan, P. I. (1993) Damage of Fiber-Reinforced Composite Materials with Micromechanical Characterization, International Journal of Solids and Structures, Vol. 30, No. 20, pp. 2757-2778.

Voyiadjis, G. Z. and Kattan, P. I. (1996) On the Symmetrization of the Effective Stress Tensor in Continuum Damage Mechanics, Journal of the Mechanical Behavior of Materials, Vol. 7, No. 2, pp. 139-165.

Voyiadjis, G. Z. and Kattan, P. I. (1999) Advances in Damage Mechanics: Metals and Metal Matrix Composites, Elsevier, The Netherlands.

Voyiadjis, G. Z. and Kiousis, P. (1987) Stress Rate and the Langrangian Formulation of the Finite-Strain Plasticity for a von Mises Kinematic Hardening Model, International Journal of Solids and Structures, Vol. 23, No. 1, pp. 95-109.

Voyiadjis, G. Z. and Mohammad, L. (1976) Rate Equations for Viscoplastic Materials Subjected to Finite Strain, International Journal of Solids and Structures, Vol. 12, pp. 81-97.

Voyiadjis, G. Z. and Park, T. (1995a) Anisotropic Damage of Fiber Reinforced MMC Using An Overall Damage Analysis, Journal of Engineering Mechanics, Vol. 121, No. 11, pp. 1209-1217.

Voyiadjis, G. Z. and Park, T. (1995b) Local and Interfacial Damage Analysis of Metal Matrix Composites, International Journal of Engineering Science, Vol. 33, No. 11, pp. 1595-1621.

Voyiadjis, G. Z. and Park, T. (1996a) Anisotropic Damage Effect Tensors for the Symmetrization of the Effective Stress Tensor, Journal of Applied Mechanics, Vol. 64, No. 1, pp. 107-110.

Voyiadjis, G. Z. and Park, T. (1996b) Anisotropic Damage for the Characterization of the Onset of Macro-Crack Initiation in Metals, International Journal of Damage Mechanics, Vol. 5, pp. 68-92.

Voyiadjis, G. Z. and Park, T. (1997) Anisotropic Damage Effect Tensors for the Symmetrization of the Effective Stress Tensor, Journal of Applied Mechanics, Vol. 5, pp. 106-110.

Voyiadjis, G. and Park, T. (1999) Kinematics of Damage for Finite-Strain Plasticity, International Journal of Engineering Science, Vol?, pp. 1-28.

Voyiadjis, G. Z. and Sivakumar, S. M. (1991) A robust kinematic hardening rule with ratchetting effects: Part I. Theoretical formulation, Acta Mechanica Vol. 90, pp. 105-123.

Voyiadjis, G. Z. and Sivakumar, S. M. (1992) A Finite Strain and Rate Dependent Cyclic Plasticity Model for Metallics, In C. Tenddosiu, J. L. Raphanel, and F. Sidoroff editors, Proceedings of the International Seminar MECAMAT91, Fountainbleau, France, on Large Plastic Deformation: Fundamental Aspects and Applications to Metal Forming, pp. 353-360.

Voyiadjis, G. Z. and Sivakumar, S. M. (1994) A robust kinematic hardening rule with ratchetting effects: Part n. Application to nonproportionalloading cases, Acta Mechanica Vol. 107, pp. 117-136.

Voyiadjis, G. Z., Thiagarajan, G. and Petarkis, E. (1995) Constitutive modelling for granular media using an anisotropic distortional model. Acta Mechanica Vol. 110, pp. 151-171.

Voyiadjis, G. Z. and Venson, A. R. (1995) Experimental Damage Investigation of a SiC- Ti Alumini de Metal Matrix Composite, International Journal of Damage Mechanics, Vol. 4, No. 4, pp. 338361.

Wang, C. C. (1970) A New Representation Theorem for Isotropic Functions: An Answer to Professor G. F. Smith's Criticism of my Papers on Representations of Isotropic Functions, Part 2, Arch. Rational Mech. Anal., Vol. 36, pp. 198-223.

Werde, R. (1972) Introduction to Vector and Tensor Analysis, Dover, New York, USA.

Wilt, T. and Arnold, S. (1994) A Coupled/Uncoupled Deformation and Fatigue Damage Algorithm Utilizing the Finite Element Method, NASA TM 106526, NASA, Lewis Research Center, Cleveland, OH.

Zbib, H. M. (1993) On the Mechanics of Large Inelastic Deformations: Kinematics and Constitutive Modeling, Acta Mechania, Vol. 96, pp. 119-138.

Ziegler, H. (1959) A Modification of Prager's Hardening Rule, Quarterly of Applied Mathematics, Vol. 17, pp. 55-65.

Index

Printed in the United States
by Baker & Taylor Publisher Services